Andreas Unger

# Refractive index sensing with localized plasmonic resonances

Andreas Unger

# Refractive index sensing with localized plasmonic resonances

Theoretical description and experimental verification

Südwestdeutscher Verlag für Hochschulschriften

**Impressum/Imprint (nur für Deutschland/ only for Germany)**
Bibliografische Information der Deutschen Nationalbibliothek: Die Deutsche Nationalbibliothek verzeichnet diese Publikation in der Deutschen Nationalbibliografie; detaillierte bibliografische Daten sind im Internet über http://dnb.d-nb.de abrufbar.
 Alle in diesem Buch genannten Marken und Produktnamen unterliegen warenzeichen-, marken- oder patentrechtlichem Schutz bzw. sind Warenzeichen oder eingetragene Warenzeichen der jeweiligen Inhaber. Die Wiedergabe von Marken, Produktnamen, Gebrauchsnamen, Handelsnamen, Warenbezeichnungen u.s.w. in diesem Werk berechtigt auch ohne besondere Kennzeichnung nicht zu der Annahme, dass solche Namen im Sinne der Warenzeichen- und Markenschutzgesetzgebung als frei zu betrachten wären und daher von jedermann benutzt werden dürften.

Verlag: Südwestdeutscher Verlag für Hochschulschriften Aktiengesellschaft & Co. KG
Dudweiler Landstr. 99, 66123 Saarbrücken, Deutschland
Telefon +49 681 37 20 271-1, Telefax +49 681 37 20 271-0
Email: info@svh-verlag.de
Zugl.: Mannheim, Universität Mannheim, Dissertation, 2010

Herstellung in Deutschland:
Schaltungsdienst Lange o.H.G., Berlin
Books on Demand GmbH, Norderstedt
Reha GmbH, Saarbrücken
Amazon Distribution GmbH, Leipzig
**ISBN: 978-3-8381-1904-5**

**Imprint (only for USA, GB)**
Bibliographic information published by the Deutsche Nationalbibliothek: The Deutsche Nationalbibliothek lists this publication in the Deutsche Nationalbibliografie; detailed bibliographic data are available in the Internet at http://dnb.d-nb.de.
 Any brand names and product names mentioned in this book are subject to trademark, brand or patent protection and are trademarks or registered trademarks of their respective holders. The use of brand names, product names, common names, trade names, product descriptions etc. even without a particular marking in this works is in no way to be construed to mean that such names may be regarded as unrestricted in respect of trademark and brand protection legislation and could thus be used by anyone.

Publisher: Südwestdeutscher Verlag für Hochschulschriften Aktiengesellschaft & Co. KG
Dudweiler Landstr. 99, 66123 Saarbrücken, Germany
Phone +49 681 37 20 271-1, Fax +49 681 37 20 271-0
Email: info@svh-verlag.de

Printed in the U.S.A.
Printed in the U.K. by (see last page)
**ISBN: 978-3-8381-1904-5**

Copyright © 2010 by the author and Südwestdeutscher Verlag für Hochschulschriften Aktiengesellschaft & Co. KG and licensors
All rights reserved. Saarbrücken 2010

Diese Arbeit entstand am Max-Planck-Institut für Polymerforschung Mainz im Arbeitskreis Materialwissenschaften unter Projektleiter Dr. Max Kreiter.

Es kommt mir vor, als hielte ich in meinen Händen den Anfang vieler dieser Wege, aber wie schwer ist es, im Nebel der Zukunft ihre Richtung, ihre Fortsetzung und ihr Ende zu erkennen! In diesem Nebel bewegen und ballen sich Naturgewalten, die der Mensch noch nicht gebändigt hat, die sich noch in keinen Plan fügen und deren Gesetze noch kein Mathematiker erforscht hat. Und unser Marsch inmitten dieser Gewalten hat eine eigene Schönheit.

Anton Semjonowitsch Makarenko

## Abstract:

In this thesis the sensing properties of plasmonic resonators for changes in the surrounding refractive index are investigated. A self-consistent and general sensing theory is developed. This theory connects the electrodynamic properties of plasmonic resonators like resonance wavelength and electric field distribution to the sensitivity for refractive index changes. A figure of merit (FOM) is derived which includes the effects of noise and in its general form directly states if a certain change in refractive index will be measurable or not.

For the FOM in the quasi-static limit absolute bounds and scalings are derived. These bounds are based on the localization of electromagnetic energy for which analytic expressions were known before. The important result of the quasi-static considerations is that the sensitivity is determined completely by the choice of material and resonance wavelength for refractive index changes that cover the whole sensing volume while for smaller analytes the energy confinement to the analyte volume is important.

To confirm the developed theory numerical calculations and an experiment with crescent shaped plasmonic resonators is carried out and good agreement is found. In this experimental verification, local refractive index changes were introduced close to the crescent shaped particles and their resonance wavelength change was measured. As a model analyte polystyrene colloids were used and manipulated with an atomic force microscope. This approach leads to a very defined and controllable model system which allowed the theoretical predictions to be verified without parasitic effects. The proposed theoretical model predicts the measured wavelength changes with high accuracy and allows to extrapolate the result to the response of the resonator to the binding of a single molecule to its surface. From the theory together with the experiment it follows, that single molecule sensitivity will be possible by increasing the signal to noise ratio of the measurement.

## Zusammenfassung:

In dieser Arbeit wurden das Verhalten von plasmonischen Resonatoren bei Veränderung des Brechungsindex der Umgebung untersucht. Für dieses Verhalten wurde eine theoretische Beschreibung entwickelt. Diese Theorie verknüpft die elektrodynamischen Eigenschaften plasmonischer Resonatoren, wie zum Beispiel das Feldprofil und die Resonanzwellenlänge, mit den sensorischen Eigenschaften der Resonatoren. Eine "Figure of Merit (FOM)" welche die Einflüsse der Messung, wie etwa Rauschen, beinhaltet wurde entwickelt. Diese kann in ihrer allgemeinen Form benutzt werden um Aussagen darüber zu treffen, ob ein bestimmtes Ereignis detektierbar ist oder nicht.

Im quasistatischen Limit können für diese FOM Grenzwerte und allgemeine Skalierungsfunktionen analytisch entwickelt werden. Ein wichtiges Ergebnis dieser analytischen Theorie ist, das für Brechungsindexänderungen die das gesamte Modenvolumen des Resonators ausfüllen, die Wahl des Resonatormaterials und der Resonanzwellenlänge die FOM und damit die Detektionseigenschaften vollständig festlegt. Für Analyte die kleiner als das Modenvolumen sind, kann die Energiekonzentration auf den Analyt optimiert werden.

Um die entwickelte Theorie zu überprüfen wurden Experimente mit hörnchenförmigen plasmonischen Resonatoren durchgeführt. In diesen Experimenten wurden kleine Variationen des lokalen Brechungsindex erzeugt, und die Frequenzverschiebung der plasmonischen Resonanz gemessen. Als Modelanalyte dienten hierbei kleine Polystryrolpartikel welche mit dem Rasterkraftmikroskop an die plasmonischen Resonatoren angebracht wurden. Es wurde so ein sehr exakt definiertes Modellsystem geschaffen, welches geeignet ist die theoretischen Vorraussagen zu überprüfen. Die Vorhersagen des entwickelten theoretischen Modells werden durch die Experimente bestätigt. Die berechneten Wellenlängenverschiebungen stimmen sehr gut mit den gemessenen überein. Das theoretischem Modell liefert zusammen mit den Messungen eine Aussage darüber ob die Detektion der Anbindung einzelner Moleküle an die Resonatoroberfläche möglich ist. Kernaussage ist, das die Detektion einzelner Moleküle bei Optimierung des Signal-zu-Rausch Verhältnisses der Messung möglich ist.

# Table of contents

1. Introduction ........................................................................................................... 1
2. Theoretical foundations ...................................................................................... 3
   2.1. Localized Surface plasmons ........................................................................ 4
      2.1.1. Optical properties of a small metal sphere ......................................... 4
      2.1.2. General properties of quasi-static plasmons ....................................... 8
   2.2. Optical properties of metals ........................................................................ 9
      2.2.1. The Drude model ................................................................................. 9
      2.2.2. Finite size effects ............................................................................... 12
   2.3. Basic sensor topology ................................................................................ 12
3. Methods .............................................................................................................. 14
   3.1. Confocal dark field microscopy ................................................................ 14
      3.1.1. Basics ................................................................................................. 14
      3.1.2. Optimization of the mechanical design ............................................ 15
      3.1.3. Spectroscopy ...................................................................................... 19
   3.2. FDTD ......................................................................................................... 24
   3.3. FEM ........................................................................................................... 31
4. Sensing theory .................................................................................................... 32
   4.1. Derivation of a Figure of Merit for the sensitivity ................................... 33
   4.2. Measurement uncertainty .......................................................................... 33
   4.3. An analytical expression for the peak shift .............................................. 36
      4.3.1. Weakly radiating systems .................................................................. 36
      4.3.2. Strongly radiating systems ................................................................ 37
      4.3.3. Expression in terms of refractive index and interpretation .............. 38
      4.3.4. Generalization for strongly dispersive resonators ............................ 38
      4.3.5. Discussion of the quasi-static case .................................................... 39
      4.3.6. Comparison with Literature ............................................................... 40
      4.3.7. An example of frequency shifts of a quasi-static resonator ............. 41
      4.3.8. Deviations from the quasi-static behaviour. ..................................... 43
   4.4. The analytical expression for the FOM .................................................... 46
   4.5. The FOM for different sensor concepts .................................................... 48
   4.6. The FOM for different types of analytes .................................................. 50
   4.7. Scaling of the FOM ................................................................................... 51
5. Characterization of the crescent shaped particles ............................................ 55
   5.1. Introduction ............................................................................................... 55

5.2. Parameterization of the crescent model ............................................................ 56
5.3. Simple Model .................................................................................................. 57
5.4. Simulation ....................................................................................................... 61
5.5. Numerical simulation of the bend Rod model ................................................ 61
5.6. Sensing properties of a simple crescent shaped resonator .............................. 65
    5.6.1. Bare resonator ......................................................................................... 65
    5.6.2. Response to attached layers .................................................................... 66
    5.6.3. Response to attached colloids ................................................................. 69
5.7. Scaling of the sensing properties .................................................................... 72
5.8. Summary and Conclusions .............................................................................. 76
6. Local response experiment ..................................................................................... 77
  6.1. Fabrication and Experiment ............................................................................ 78
    6.1.1. Fabrication of the crescent samples ........................................................ 78
    6.1.2. Determination of the optical response .................................................... 79
    6.1.3. Measurement of the spectra .................................................................... 80
  6.2. Results and Discussion ................................................................................... 83
  6.3. Extrapolation to Single molecules .................................................................. 88
7. Dense Arrays of crescent shaped resonators .......................................................... 90
  7.1. Coupling Effects ............................................................................................. 90
  7.2. Setup of the simulations ................................................................................. 91
  7.3. Effect of crescents distance ............................................................................ 92
  7.4. Effect of crescents rotation ............................................................................. 93
  7.5. Effect of crescents deformation ..................................................................... 95
  7.6. Experimental results and discussion ............................................................... 95
8. Summary ................................................................................................................. 98
Appendix A Measurement uncertainty for a Lorentzian peak ................................... 100
Appendix B meep code .............................................................................................. 102
Appendix C Table of figures ...................................................................................... 122
Appendix D List of Symbols ...................................................................................... 125
Literature .................................................................................................................... 128
List of Publications .................................................................................................... 131

# 1. Introduction

There is an ever growing demand for the precise detection of molecules. For example the progress in the understanding of diseases on a molecular level will lead to new methods of treatments of diseases which are currently not possible. This will require the ability to detect the presence of only very few specific molecules. Other fields which require high sensitivity are gas sensing for safety systems and measurements of chemical or physical quantities like pH-values, temperatures or phase transitions. The measurement of such quantities usually needs a transduction process, where the primary quantity of concern is transformed into a physically measurable quantity like an electric voltage. One of the main transducer principles is to translate the measurement signal into an optical signal, in many cases a spectroscopical one. This approach is known to be very robust and flexible, and many sensors have been designed based on this principle. Here two of the most important classes are the surface-plasmon-resonance (SPR) based sensors used widely in biochemical research [1] and gas sensing and the surface enhanced infrared absorption spectroscopic methods which can be used for the precise measurement of the composition of chemical molecules.

Another relatively new method is to use small metal particles as transducers. The basic principle here is that the particles are so small that an incoming oscillating electrical field can penetrate them and shake the free electrons inside the particle hence the particle becomes an oscillator. This oscillator will show a resonance at visible or near infrared wavelength which will show up in extinction or scattering spectra as an extinction peak. This resonance is called localized surface plasmon resonance (LSPR). The spectral position of this peak is highly dependent on the type of metal, the shape of the resonator, and - what makes it interesting for sensing - on the surrounding of the metal particle. The LSPR is known since a long time (books on the optical properties of small metal particles are for example [2, 3] ) but only recently its high potential for sensing applications has been recognized. The first use of colloidal gold particles as sensors was 1998 [4] but rapid development of the field followed. A detailed description of some commercially available sensing platforms can be found in the PhD thesis of Raschke [5]. When it comes to high sensitivity applications where detection of only few molecules is required LSPRs have a number of very favourable properties: The modal volume offered by LSPRs is in the order of only $1000 nm^3$ which means that only few molecules will fit into this volume. This makes them the natural choice for such applications. For the application of LSPRs usually noble metal particles like gold particles are employed. These have a high chemical inertness and it is easy and well known how to modify their surface to allow only for specific binding of molecules. The metal particles are very strong absorbers and scatterers so that the observation of a single resonator, which is much smaller than the diffraction limit, is possible in a standard microscope.

Consequently, to come down to the few molecule sensitivity techniques were developed to monitor spectra of single particles in real time [6, 7]. In the last ten years numerous experiments were carried out which showed the sensing capabilities of single resonators for changes in refractive index and for binding of biomolecules [8-10]. In these experiments the refractive index of a thin layer around the resonator or in the whole surrounding of the resonator was changed and the spectral peak shift of the resonance was measured. Different particle shapes where used and the sensitivities of the different shapes were compared.

When it comes to a theoretical description of the sensing process the available studies are very limited. Miller and Lazarides predicted the sensitivity of ellipsoids in the quasi-static limit [11] on changes of bulk refractive index. Others simply declared a figure of merit for sensitivity optimization from intuitive arguments rather than from the underlying physics [12]. No self-consistent and general sensing theory was available at the time this thesis was started. Of course, it is possible for a given plasmonic resonator to numerically determine its response and even do optimization studies. Still it is desirable to have an analytical theory at hand for several reasons: Firstly from such a theory absolute limits of detection could be derived which show the underlying physical limitations. Secondly an analytic theory can guide the optimization procedure to reach these limits by clarifying what has to be optimized. The question what has to be optimized sounds trivial but it is not. For example when different parameters are to be optimized it may occur that one parameter can only give the double sensitivity before hitting the physical limit, while optimizing another could improve the sensitivity 100 times. Clearly in this case it makes no sense to optimize the first parameter but the second. Thirdly, the present situation for numerical calculations is that 3D computations of spectra of even the smallest systems still take computational times in the order of hours, making it very difficult to do algorithmic optimization. Therefore an analytical theory would here help to reduce optimization times by providing good starting guesses for optimization.

This work is intended to close the gap and provide a general sensing theory. A theoretical description of the sensing process is developed. With this description it is possible to accurately predict sensing capabilities of general plasmonic resonators. A specialization of the developed approach to the quasi-static limit combined with energy considerations leads to absolute limits on the sensitivity for small resonators. The sensitivity model is then combined with a statistical description of the detection process to yield a general figure of merit for the resonator as well as the optical instrument.

Based on the theoretical description crescent shaped resonators are characterized in terms of their sensing properties. It is shown that this kind of resonator has very interesting and promising sensing capabilities. Crescent shaped plasmonic resonators were first introduced by Shumaker-Perry and Rochholz [13, 14]. They can be very easily fabricated in a lithographic process and their plasmonic resonances are tunable over a wide spectral range from the lower end of the visible into the middle infrared. Additionally, they possess very sharp features where high field localization occurs and have multiple resonances. The first feature decreases the effective modal volume which is interesting for single molecule sensing while the second allows in principle to determine more than one quantity of an analyte simultaneously.

With the theoretical description at hand an experiment is designed and carried out to show the response of crescent shaped resonators to strongly localized changes in refractive index. Because single molecule experiments are difficult to interpret due to parasitic effects a much more defined approach is chosen here: Single molecules are mimicked by small polystyrene colloids which can be manipulated in a defined way by atomic force microscopy. The experimental results are then compared to the theory. Very good agreement was found. These results confirm the theoretical approach and show how to optimize the system to bridge the remaining gap to true single molecule sensitivity.

The thesis is organized as follows: At first an introduction to the theoretical basics is given. In the next chapter the used numerical and experimental methods are introduced. The experiments carried out later require an absolute wavelength precision of the measured spectra better then 1 nm. In this chapter it is also reported how this precision was achieved. Then the sensing theory is developed and compared to several numerical simulations. Applicability and limits of this theory are discussed. Then the theory is specialized to the quasi-static regime. It is found that in this regime scaling properties in general and the sensitivity in special cases can be predicted without any numerical calculations.

Next the response of crescent shaped resonators is calculated numerically as well as with simple approximate models. A full characterization of the sensing capabilities is carried out and compared to already available experimental results. Next an experiment for the response of crescent shaped resonators is carried out and described. Based on the experimental results an extrapolation to the response on single molecules is done. It follows a chapter on additional effects of ordered crescent arrays and finally a summary.

## 2. Theoretical foundations

This chapter is intended as a short overview to the theory of localized surface plasmon resonators and molecular sensing. It is not intended to fully describe the theory involved but only very basic aspects important for this thesis will be reviewed. Where available, citations to the original literature are given for further reading. The chapter starts with the description of the optical properties of a metal sphere, one of the simplest geometries which where a localized plasmon can be excited. Then very useful general properties of LSPR's are introduced, based on a quasi-static analysis. Third the chapter reviews the dielectric properties of metals described in a classical, non quantum mechanical fashion, which is usually exact enough in practice. Last a description of a typical molecular sensor is given and shown how its function is divided into partial functions.

## 2.1. Localized Surface plasmons

### 2.1.1. Optical properties of a small metal sphere

A small metal sphere is the simplest geometry which shows a localized plasmon resonance. It has the advantage that the theoretical solution of it's response to an incoming plane wave is long known and was already found by Mie in 1908 [15]. While it may not be the best geometry as a sensor, the existence of an analytical solution allows to compare theoretical predictions and experiments and to understand qualitatively also the resonances of more complex geometries. Consequently the first experiments on localized plasmons and also the use of LSPR's as sensors where carried out with small gold spheres [4]. This chapter is intended to show the basic properties of a LSPR.

A time harmonic electromagnetic field in a homogeneous space fulfils the wave equations[16]:

$$\Delta^2 \vec{E} + k^2 \vec{E} = 0 \quad \Delta^2 \vec{H} + k^2 \vec{H} = 0 \tag{1}$$

Where E and H are the electric and magnetic fields and k is the wave number in the medium

$$k^2 = \frac{\omega^2}{c^2} = \omega \varepsilon_0 \mu_0 \varepsilon \mu \tag{2}$$

$\omega$ is the angular frequency of the time harmonic wave c is the speed of light in the medium, $\varepsilon$ is the permittivity of the medium and $\mu$ is the permeability of the medium which is usually one at optical frequencies and is therefore neglected here. The electric and magnetic fields are in the absence of charges divergence free and connected via Faradays and Amperes law. The wave equations (1) can be solved exactly only in very few cases, typically only when a coordinate system can be found where the physical boundaries of the system conform to constant coordinate planes and the wave equation becomes separable into a set of ordinary differential equations. This is the case for spheres or infinite cylinders. In the case of spheres the result is the expansion of the electric and magnetic fields in the form

$$\vec{E} = \sum_{n=1}^{\infty} E_n \left( i a_n \vec{N}_{e1n}^{(1,3)} - b_n \vec{M}_{o1n}^{(1,3)} \right)$$
$$\vec{H} = \frac{k}{\omega} \sum_{n=1}^{\infty} E_n \left( i a_n \vec{N}_{o1n}^{(1,3)} - b_n \vec{M}_{e1n}^{(1,3)} \right) \quad (3)$$

$E_n$ is $i^n E_0 (2n+1)/n(n+1)$, $E_0$ the incoming field strengths and the N and M functions are known as the vector spherical harmonics. The $a_n$ and $b_n$ are the expansion coefficients and determined by matching the fields at the boundary of the sphere. Since the whole theory is lengthy it will not be repeated here. A modern derivation can be found in [2]. For experiments the observable quantity is the scattered or absorbed light power. These quantities are usually described by the scattering and extinction cross sections $\sigma$ defined as the areas which give when multiplied with the incoming intensity of a plane wave $I_{In}$ the total scattered or absorbed power P.

$$P_{Scat/Abs} = \sigma_{Scat/Abs} I_{In} \quad (4)$$

Often the scattering or absorption cross sections are described with the dimensionless efficiency Q which when multiplied with the area of the scatterer projected perpendicular to the power flow of the incoming plane wave give the cross section $\sigma$.

$$\sigma_{Scat/Abs} = Q_{Scat/Abs} A_{\perp} \quad (5)$$

Another quantity used is the extinction cross $\sigma_{Ext}$ section defined as the sum of scattering and absorption cross section.

$$\sigma_{Ext} = \sigma_{Scat} + \sigma_{Abs} \quad (6)$$

From the Mie theory the cross sections are given by

$$\sigma_{Scat} = \frac{2\pi}{k^2} \sum_n (2n+1) \left( |a_n|^2 + |b_n|^2 \right)$$
$$\sigma_{Ext} = \frac{2\pi}{k^2} \sum_n (2n+1) \operatorname{Re}[a_n + b_n] \quad (7)$$

When a scatterer gets very small at some point the retardation over its whole volume will become negligible. This requires that the phase shift $\varphi$ over the particle is much smaller then $2\pi$

$$\varphi = \frac{2\pi}{\lambda_0} na \ll 2\pi \quad (8)$$

When this condition is satisfied, all parts of the scatterer will see the incoming field in the same phase and field strength. In this case only the first term of the expansion (3) is important and all others are negligibly small. In this case the scattering and extinction cross section simplify to

$$\sigma_{Ext} = 3kV \, \text{Im}\left[\frac{n^2 - n_0^2}{n^2 + 2n_0^2}\right]$$

$$\sigma_{Scat} = \frac{3k^4 V^2}{2\pi}\left|\frac{n^2 - n_0^2}{n^2 + 2n_0^2}\right|^2 \qquad (9)$$

Where n is the refractive index of the sphere and $n_0$ the refractive index of the ambient. The extinction for such a small scatter is dominated by absorption and is proportional to the volume V of the sphere while scattering is proportional to the square of the volume. The same result can be derived in an electrostatic approximation. In this case one assumes from the beginning that retardation is zero and all fields have constant phase. In this case the rotation of the electric field is zero and it can be derived from a scalar potential

$$\vec{E} = \nabla \Phi \qquad (10)$$

And the problem reduces to solving the Poisson equation.

$$\Delta \Phi = 0 \qquad (11)$$

The solution of this equation for the sphere leads to the same expressions for the cross sections then the small particle Mie limit. This shows the connection between the quasi-static description and an analysis which takes into account retardation. That the quasi-static case is the small particle limit of the general description will be important in following chapters. Another outcome of the quasi-static description is that the sphere can actually be described as a dipole when one is far enough away from the surface. One can assign it a dipole moment p and a polarizability $\alpha$ which is defined as the linear response to an incoming electrical field $E_0$[17]

$$\vec{p} = n_0^2 \alpha \vec{E}_0 \qquad (12)$$

The polarizability of a small sphere turns out to be

$$\alpha = 3V \frac{n^2 - n_0^2}{n^2 + 2n_0^2} \qquad (13)$$

which allows to write the cross sections in terms of $\alpha$

$$\sigma_{Ext} = \frac{k}{\varepsilon_0} \text{Im}[\alpha]$$

$$\sigma_{Scat} = \frac{k^4}{6\pi \varepsilon_0^2} |\alpha|^2 \qquad (14)$$

These are the basic equations for the scattering properties of a small dielectric sphere. Their implications will now be discussed. The observed scattered intensity of a small sphere is the product of two parts. One is proportional to $k^4$ or $\lambda^{-4}$ and dominates the spectral response when the dispersion is weak. This effect is known as Raleigh scattering named after Lord Raleigh who used this theoretical model to first explain the blue colour of the sky. It implicates that small wavelengths are much stronger scattered then longer ones, leading to the blue appearance of a turbulent or dusty media which contains very small particles when observed not directly towards the light source. A light source looks red when observed trough such a medium. The second part is material dependent and is $\frac{n^2 - n_0^2}{n^2 + 2n_0^2}$. When the denominator of this expression tends to zero the scattering and absorption will get very large. This leads to a resonance condition

$$n^2 = \varepsilon = -2n_0^2 \qquad (15)$$

For a normal dielectric material this cannot be fulfilled since dielectrics have an almost pure real refractive index and therefore positive permittivity. For metals below their plasma frequency however the permittivity is negative. When the incident light approaches the plasma frequency at some wavelength the condition (15) will be fulfilled leading to an additional resonance peak in scattering and absorption spectra. This resonance is known as the localized surface plasmon resonance (LSPR). Figure 1 shows an extinction spectrum of a 40nm in diameter gold sphere calculated with the Mie theory. Clearly the plasmon resonance superimposed on the Raleigh scattering background can be observed.

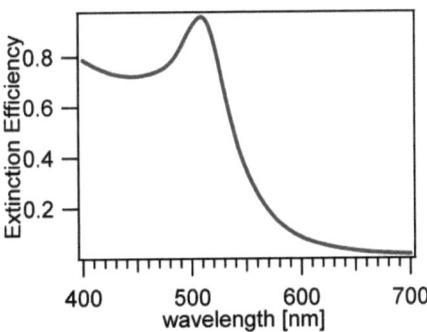

**Figure 1: Extinction for a 40nm in diameter gold sphere.**

From (15) a very useful property can be found immediately. The permittivity at which the resonance occurs is dependent on the refractive index of the surrounding medium $n_0$. Since metals are dispersive a change of the refractive index of the surrounding will shift the resonance to a different wavelength. This shows basically the sensing capabilities of plasmonic resonators in a very simple case. The expression for the polarizability of a sphere can be generalized to ellipsoidal particles illuminated with the polarization along one of the principal axes [2]:

$$\alpha = 3\pi V n_0^2 \frac{n^2 - n_0^2}{n_0^2 + L(n^2 - n_0^2)} \quad (16)$$

Here L is the depolarization factor which depends on the aspect ratio of the ellipsoid. The case of a sphere is contained as a special case with L =1/3.

## 2.1.2. General properties of quasi-static plasmons

In the previous chapter the properties of a special plasmonic resonance were shown. Unfortunately this is one of the few cases which can be calculated analytically. The question was therefore if one could find general properties which all plasmonic resonators have to fulfil. Wang and Shen [18] derived such general properties based on an energy consideration in the quasi-static limit. These properties turned out to be very useful for this work and will therefore be repeated here. The first property is that for any local mode of a quasi-static structure a constitutive relation for the electric field E holds:

$$\int_\Omega \varepsilon \vec{E}^2 dV = 0 \quad (17)$$

Here ε is a function of position and Ω is all space. This relation follows purely from the assumption that the system is quasi-static and hence E can be derived from a scalar potential (see equation (10)). For a system composed of two materials where one is a metal with $\varepsilon_M$ and the other is a dielectric with $\varepsilon_D$ this lead to a resonance condition equivalent to (15) but for general systems:

$$\varepsilon_M = - \int_{Dielectric} \varepsilon_D \vec{E}^2 dV \Big/ \int_{Metal} \vec{E}^2 dV \quad (18)$$

Based on this expression and the analytical properties of the dielectric function it was further shown that the energy stored in the metal $U_M$ is always bigger then in the dielectric $U_D$ and the ratio is given by

$$U_M / U_D = \frac{d(\omega \varepsilon_M')}{d\omega} \Big/ -\varepsilon_M' \geq 1 \quad (19)$$

Where $\varepsilon'_M$ is the real part of the metal permittivity. It can be seen that this ratio is completely defined by the material properties of the metal alone, and is independent of the microstructure of the system. Another important quantity of a resonance is the quality factor Q which is defined as the ratio of energy stored in the resonant field and the energy lost per oscillation period of the resonance. Q is a measure of the resonators ability to store energy and is connected to the line width Γ of a resonance via [16]

$$Q = \omega/\Gamma \qquad (20)$$

Here ω is the resonance frequency and Γ the full line width at half maximum intensity (FWHM). It turns out that Q in the quasi-static limit is again completely determined by the material properties of the resonator and given by

$$Q_{static} = \frac{\omega d\varepsilon'_M/d\omega}{2\varepsilon''_M} \qquad (21)$$

Here $\varepsilon''_M$ is the imaginary part of the permittivity of the metal. This results show that many properties of a plasmonic resonance like field and energy distribution and quality of the resonance are determined just by the type of metal used and the resonance wavelength. One would expect that such strict limits have an influence on other properties like sensitivity to the environment as well and it will be shown later that indeed bounds on this sensitivity can be derived from these results.

## 2.2. Optical properties of metals

As plasmonic effects require metals, this chapter will provide a short review of the dielectric properties of metals at optical frequencies. A remarkable fact is, that this properties are explainable to a huge extend in a pure classical model. In this model the free electrons in a metal are simply treated as classical free particles which are driven by the externally applied field and satisfy a classical equation of motion. This model has been first proposed by Drude [19].

### 2.2.1. The Drude model

The Drude model considers the movement of a classical electron in an ionic background under the influence of an externally applied field. The equation of motion of such an electron is

$$m_e \ddot{x}(t) + \frac{1}{\tau}\dot{x}(t) = -qE_0 e^{-i\omega t} \qquad (22)$$

Here x is the displacement of the electron from its equilibrium position, $m_e$ is the electron mass, q is the magnitude of the electron charge, $E_0$ the magnitude of the applied external field and ω it's angular frequency. The first term describes the acceleration of the electron in an electric field while the second one ensures that under a DC field an electron will reach a constant drift velocity which is the observed behavior. This velocity is actually an average one, averaged over many electrons and describes the fact that microscopically after a mean relaxation time $\tau$ the electron will get scattered into a random direction. A particular solution to (22) is:

$$x(t) = x_0 e^{-i\omega t} = E_0 e^{-i\omega t} \frac{q/m}{\omega^2 + i\omega/\tau} \tag{23}$$

The Polarization P is given then by

$$P = \varepsilon_{Ion}\varepsilon_0 \chi E = -\rho q x = E_0 e^{-i\omega t} \frac{\rho q^2/m}{\omega^2 + i\omega/\tau} \tag{24}$$

Here $\chi$ is the susceptibility $\varepsilon_{Ion}$ is the permittivity of the ionic background and $\rho$ the density of electrons. Solving this equation for the susceptibility gives

$$\chi = -\frac{1}{\varepsilon_{ion}} \frac{\rho q^2/(\varepsilon_0 m)}{\omega^2 + i\omega/\tau} \tag{25}$$

The nominator of this term has the dimension of a frequency squared. Its square root is called the plasma frequency $\omega_P$ because it is the oscillation frequency of the free electron plasma.

$$\omega_p = \sqrt{\rho q^2/(\varepsilon_0 m)} \tag{26}$$

The permittivity $\varepsilon_M$ of a metal finally becomes

$$\varepsilon_M = \varepsilon_{ion}(\chi+1) = \varepsilon_{ion} - \frac{\omega_p^2}{\omega^2 + i\omega/\tau} \tag{27}$$

Usually $1/\tau$ is several orders of magnitude smaller then $\omega$ in the visible or near infrared spectrum. The real part of is $\varepsilon_M$ then in good approximation

$$\varepsilon'_M \approx \varepsilon_{ion} - \frac{\omega_p^2}{\omega^2} \tag{28}$$

Thus it is negative as long as $\omega<\omega_P/\varepsilon_{Ion}$. A negative epsilon is needed to achieve plasmonic resonances so here it can be seen why metals are needed. Figure 2 shows the permittivity of silver which is a typical metal in the visible and IR and has its plasma frequency in the ultraviolet.

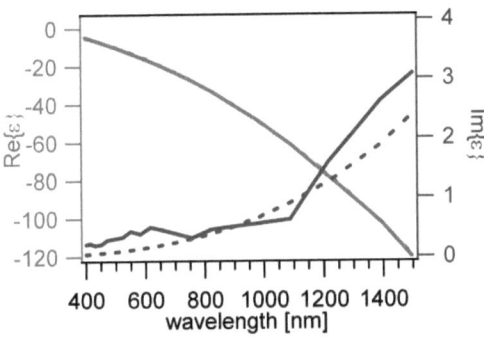

**Figure 2:** Permittivity of silver from [20] (continuous lines) and a fit with the Drude model (dashed lines). Red shows the real part while blue shows the imaginary part. The deviations in the imaginary part are because of the experimental accuracy.

The above described dielectric properties of metals results from the free electrons in the conduction band of the metal, but a metal will also have other bands with gaps between them. The energies of this band gaps lie as usual for electronic transitions in the visible and UV frequency range. When the frequencies of the transitions are reached their additional contributions to the permittivity become important. A typical metal with such interband transitions in the blue frequency range is gold. Figure 3 shows its permittivity as a function of frequency together with a fit of the Drude model. The fit was carried out in the NIR where gold is to a good approximation a Drude metal. One clearly can see strong deviations from the Drude behaviour at wavelengths smaller then 700nm.

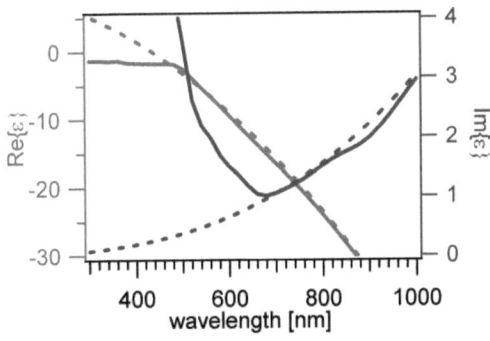

**Figure 3:** Permittivity of gold from [20] (continuous lines) and a fit with the Drude model (dashed lines). Red shows the real part while blue shows the imaginary part.

Usually these additional transitions lead to higher absorption and worsen the plasmonic behaviour of metals. It is advisable to stay out of this regions for sensing applications which will be seen clearer later in chapter 4.

### 2.2.2. Finite size effects

If a metal particle gets so small that the mean free path of the electrons gets comparable to the particle size additional damping effects become important. In a classical picture that can be understood as an additional scattering of the electrons at the surface of the metal. For very small particles this can even become the dominating loss effect. This effect can be described as an additional relaxation time for surface scattering $\tau_S$:

$$\frac{1}{\tau} = \frac{1}{\tau_{Bulk}} + \frac{1}{\tau_S} \tag{29}$$

This effect was extensively studied by Genzel, Kreibig and Vollmer[3, 21, 22]. They found out that for gold particles these effects are negligible down to particle sizes of 20nm. For smaller particles this effects are observable as a strong line broadening of the plasmonic resonance. This finite size effects are also important in bigger particles which contain small feature in this order of size. Often at small features like sharp tips the electrical fields are strongly enhanced leading to enhanced sensitivity in a small volume. The finite size effects discussed here lead to a reduction of the field strength compared to a description with the bulk permittivity and therefore limit sensitivity.

## 2.3. Basic sensor topology

In this chapter the basic sensor topology for detection of molecules is outlined. Under a bio- or chemosensor one generally understands a device which translates the presence of a molecule or a concentration of molecules into a measurable quantity like a voltage, a spectrum, a change in colour or others. It generally consists of a number of partial functions as shown schematically in figure 4.

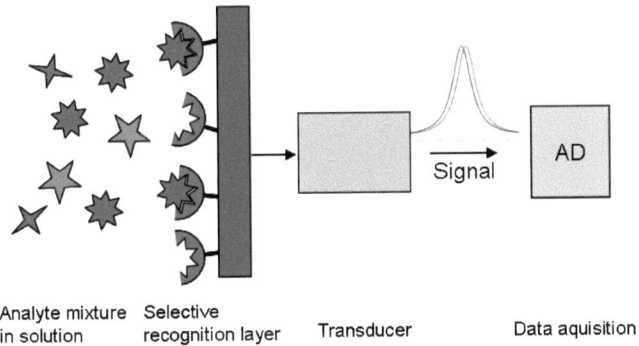

**Figure 4: Partial functions of a sensor device**

Typical the analyte is contained in a solution or buffer gas and mixed with other molecules. The first task of the sensor is to select the analyte from this mixture. For this purpose usually some kind of recognition layer is used. This layer consists of receptors which can bind to the analyte. The requirements on this layer are high chemical stability and strong selectivity for the analyte. A common problem of this layer is unspecific binding, which means that other molecules bind to it as well and lead to a wrong signal of the sensor. The recognition layer is followed by a transducer. The transducer translates the presence of molecules to the physically measurable quantities and presents the main part of the sensor. It has to fulfil a number of requirements:

- Defined and high enough sensitivity
- Easy to calibrate
- Linearity over the full measurement range
- Stable against environmental influences (chemicals, temperature)
- Low drift of its properties over time
- Reversible

There are many established transducer principles available, e.g. electrochemical, optical, and spectroscopic. When additional requirements like small sensing volume are present often optical spectroscopic methods are the methods of choice, because of the easy confinement of optical fields. As shown in the previous chapter plasmonic resonances provide extremely small sensing volumes in the order of a few ten nanometres cubed. When single molecules should be detected the linearity requirement can be abandoned for highest sensitivity. It is this range where LSPR based sensors hold the best promise and what finally this work is about.

# 3. Methods

## 3.1. Confocal dark field microscopy

### 3.1.1. Basics

Contrary to a usual bright field microscope where the whole field of view is illuminated by a light source and imaged to a detector, a confocal microscope only illuminates a single diffraction limited spot on the object and the detector is also point like. Image formation is done by scanning the laser beam over the object or scanning the object itself. figure 5 shows the basic setup for confocal microscopy in reflection. A light source is spatially filtered by an illumination pinhole and imaged onto the sample via a beam splitter and a microscope objective. The back reflected light transmits the beam splitter and is focused to a pinhole and the detector. The name confocal comes from the fact that excitation and detection light share the same focal positions contrary to a conventional microscope where the foci of illumination and imaging usually are in each others conjugated planes (e.g. Köhler illumination).

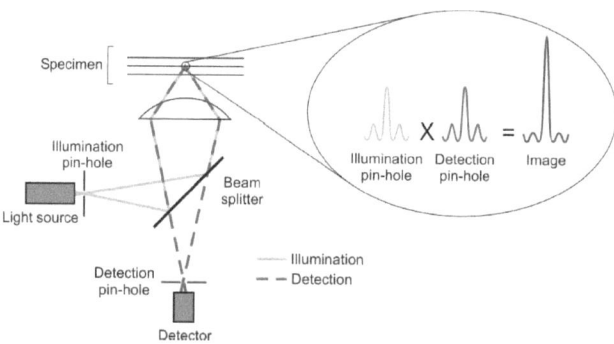

**Figure 5: Basic confocal setup for reflection microscopy [23].**

When scattering spectroscopy is to be performed it is useful to collect only the small amount of scattered light and not the full incident beam. Such a dark field mode can be easily integrated into a confocal reflection microscope by using aperture stops in the excitation and detection paths as shown in figure 6. In the excitation a blocking disc is inserted creating a ring like light beam which leads to an annular illumination when focused onto the sample. In the detection path an adjustable iris is inserted and adjusted so that the directly reflected beam is stopped. The light which is scattered into small angles by a scatterer is passed trough.

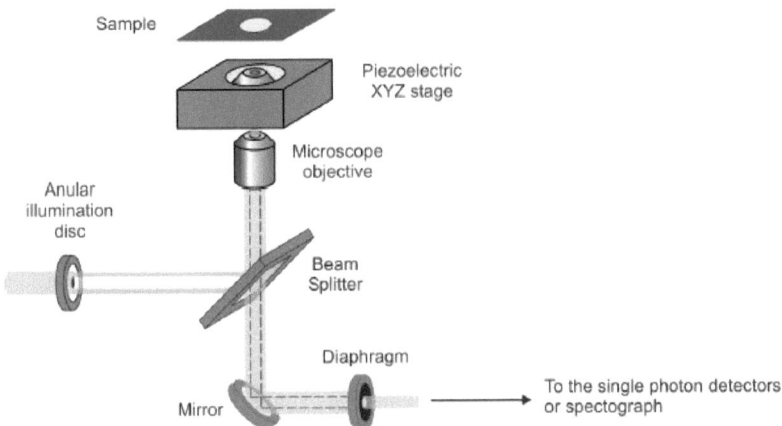

**Figure 6: Annular illumination for dark field measurements [23]**

The main advantage of the confocal dark field method is the efficient suppression of scattered light which does not come from the object plane. This suppression is reached for two reasons. First only a small focal volume is illuminated so scatterers outside this volume do not contribute to the scattered signal. Second only light from the focal volume can pass trough the detector pinhole as illustrated in figure 7.

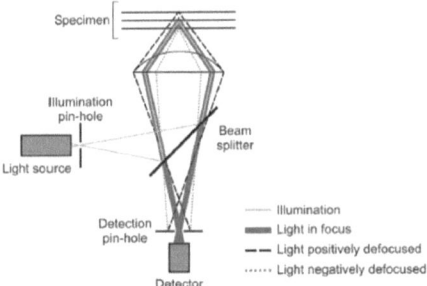

**Figure 7: Suppression of stray light in a confocal microscope [23]**

### 3.1.2. Optimization of the mechanical design

When this work was started a homebuilt confocal microscope [23] was readily available and the original intention was to just extend it for scattering spectroscopy in the NIR spectral range. However this was abandoned early as it turned out that the present mechanical design was not stable enough to reach the desired spectral precision. The main reasons for the instabilities were the use of highly floating optical elements and non-orthogonal beam adjusters. The original setup used 20cm high aluminium rods screwed to an optical table for mounting optics like lenses apertures, and mirrors. Aluminium has a thermal expansion coefficient of $23 \times 10^{-6} K^{-1}$ which leads to an expansion of such a rod of $4\mu m\ K^{-1}$. Such an expansion leads to a beam displacement of $40\mu m\ K^{-1}$ when the beam is collimated by a lens with f=100mm and propagates a meter in length (which is a typical distance trough the whole instrument). This simple example shows that in order to temperature drifts low the whole setup should be stabilized to $\Delta T<0.25K$ in order to reach drifts of less than 10 micron beam displacement (which is 10% of a typical pinhole size). Such thermal requirements are not fullfillable over days. Consequently when the first measurements were done intolerable spectral drifts were observed. Another issue is that this construction leads to vibrations.

The second problem was that beam adjustments where not decoupled in the x and y axes. The detection beam was adjusted at two mirrors which where at skew angles relative to the optical axis. When readjusting the beam in one axis it misaligned in the other making it impossible to readjust small drifts in a reproducible way.

First measurements showed that a complete redesign for stability was necessary. During the redesign process also the necessity emerged to make the system as modular as possible to extend it easily for other measurements without too much change of the basic design. To achieve these goals methods of methodical construction were used. The device was divided into partial functions and for each functional block the best design out of several identified possibilities was chosen.

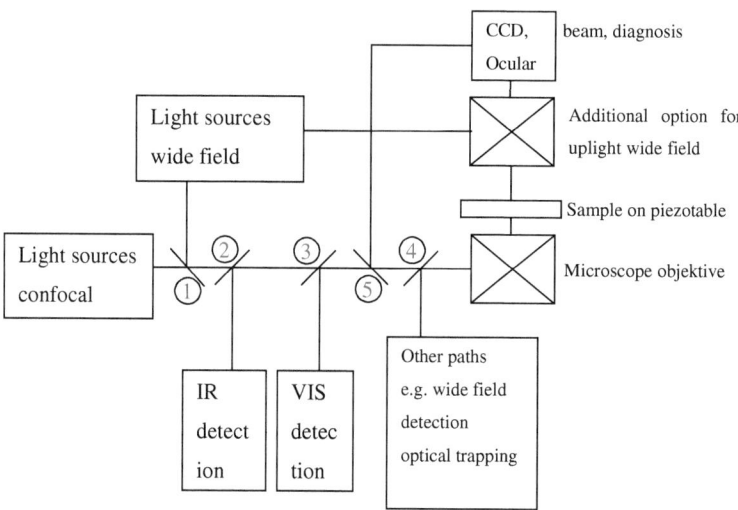

**Figure 8: Block diagram of partial functions and light signal flow of the microscope. The numbers in circles are for redirections of the beam path.**

Figure 8 shows the determined partial functions. Some of them like the light sources and the piezo table turned out to be useable as they were before and were not changed. The requirement of orthogonality made it necessary to split the beam redirection and splitting into two parts, which was one before.

As basic mechanical setup a rail based system was chosen (Owis Sys65 Owis GmbH Germany). This system comes with 65mm broad and 22mm thick rails. The beam height is 65mm over the mounting area of the rails or 40mm over the rail. This compact format gives highest stability. Optics up to 40mm diameter can be integrated. To maintain compatibility to existing detectors and light sources the rails were mounted on massive aluminium blocks of 120mm height. To minimize strains only materials with similar thermal expansion coefficients are used. The rail system has the additional advantage that it gives the required flexibility to add new beam paths without changing other parts of the setup. The whole rail is accessible for optical components and mounts for optical components come with a fixed beam height which makes it easy to plug in complete new systems. To the end of this work an additional beam path for wide field imaging and one for an ultrafast laser were installed without changing anything on the basic instrument.

It turned out that the desired wavelength range for spectroscopy (500nm to 1500nm) cannot be covered by a single set of achromatic optics. For this reason the system was built with a split beam path for VIS/NIR light (500-900nm) and one for NIR (800-1500nm).

**Figure 9: Photography of the optimized confocal microscope. The whole setup is covered in a box to shield the sensitive detectors from stray light**

Figure 9 shows a photography of the current setup and figure 10 shows schematically the arrangement of components on the table.

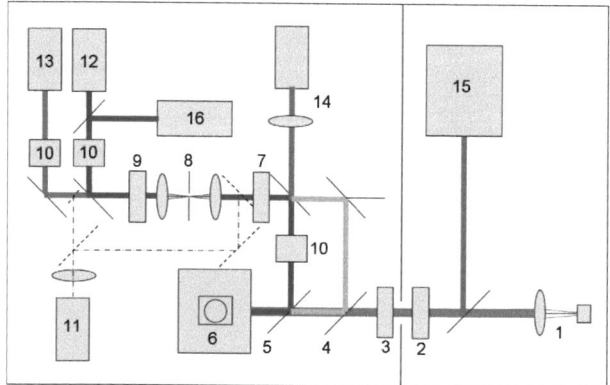

**Figure 10: The arrangement of components of the confocal microscope (roughly to scale, the size of the optical table is 2x1.5m)**

The numbered components in figure 10 are:
1. fiber coupler and collimator for fiber coupled light sources (lasers and arc lamp)
2. polarization optics
3. spatial beam shaping (e.g. for annular illumination)
4. IR beamsplitter
5. VIS/NIR beamsplitter
6. microscope table with piezo table and microscope objective
7. spatial beam shaping optics (e.g. iris aperture to block direct light)
8. confocal pinhole
9. polarization optics
10. spectral filtering
11. ocular for monitoring and diagnosis of all beam paths
12. avalanche photodiode for scanning
13. fiber coupler to grating spectrometer
14. CCD for wide field imaging
15. ultrafast laser system, free space coupled
16. NIR photomultiplier tube for NIR scanning

The coloured lines show the different beam paths. Red is the excitation path, blue is the visible detection path, orange the IR detection path, dark green the spectroscopy path and green are additional path which are not used in this work. Dashed lines are beam diagnosis path which are used for alignment of the instrument.

### 3.1.3. Spectroscopy

In this section the optical setup for single particle spectroscopy is described. The spectroscopic measurements are carried out in the dark field mode as described before. As described in chapter 4 the most important property of the optical instrument is to maximize the detected signal, because this determines the uncertainty in peak detection. The optimization must be done in a way not to sacrifice instrumental precision so a trade-off has to be found between systematic and statistical errors. The influence of several mechanical parameters of the spectroscopic setup like slit width and positioning accuracy has been investigated by Curry et. al. [24]. Their main result was that influences of the geometrical parameters like position of the particle on the spectral peak position can be minimized to less then 10pm with correct choice of the slit width of the spectrometer and feedback control of the microscope table to less than 10nm accuracy which is possible with capacitive or interferometric controlled piezo tables. These results were accounted for in the optimization of the instrument.

The final optical setup looks like follows: Light coming from a single mode fiber (SM800) is collimated with an f=100 achromatic lens which results in a roughly two times overfilling of the microscope objective (Nikon Planapo 60x 1,45NA Oil immersion). Then the light beam was directed trough a blocking disc of 5mm diameter and a broadband beam splitter plate (Linos GmbH, Germany). The reflected light was collected by the same microscope objective and after redirection from the beamsplitter the direct light was blocked by an adjustable iris aperture. The iris aperture together with the blocking disc was adjusted for maximum SNR at the spectrometer. Surprisingly this was achieved not by a complete dark field mode but with a slightly more open aperture which transmits a bit of the direct light. After the aperture light was focused onto the confocal pinhole (d = 100 µm) with a f = 80mm lens and recollimated. Then the light beam was polarized by a Glan-Thompson polarizer to select the correct polarisation of the studied resonance. A polarization of the incident beam was omitted because the input beam is 20 mm in diameter which makes low loss Glan-Thompson polarizers very expensive. Foil polarizers are no option because they have low transmittance and usually introduce serious wave front aberrations. After polarization the light beam is either focused onto an avalanche photodiode (Perkin Elmer, Optoelectronics Inc., USA) for scanning images or coupled to a multimode fibre (200 µm core with NA = 0.22) and directed to a grating spectrometer with a CCD (Andor Shamrock SR303i, Andor Technology, Belfast, Northern Ireland). For scanning, the light beam is spectrally filtered to selectively image only the relevant resonances and to reduce systematic errors (explained in the next section). The coupler from the multimode fiber to the spectrometer was optimized to maximize throughput. Best results were achieved by first collimating the light which exits the fibre with an f=40mm achromatic lens and then focusing it to the slit of the spectrometer by an f=100mm cylindrical lens. As reported by Curry et al. [24] it is essential to couple the whole airy disc to the spectrometer to reduce the effects of positioning errors on the spectrum. This is achieved by a very wide slit. The drawback of this open slit is the additional coupling of stray light to the spectrometer. By focusing the light on the slit a configuration can be found where stray light is minimized but still all light scattered by the resonator reaches the spectrometer.

### 3.1.3.1. Systematic errors and solutions to avoid them

As described in the previous section the setup the effects of positional errors on the spectrum were minimized. When measuring in the NIR region of the spectrum however it turned out that systematic errors are still large. Figure 11 shows the raw spectra of a typical crescent resonator as a function of z position (given as the control voltage of the piezo table 8µm=1V). One can see that strong spectral shifts occur by varying the focus only about 800 nm. While for itself not a big problem it turned out that it was also impossible to adjust the focus from a scanned image reliably to reach spectral shifts smaller then 1nm. When adjusting the focus for maximum intensity in the scanned image the found deviations in the spectral position where in the order of 10nm between successive scans. The reason for this behaviour turned out to be strong chromatic aberrations at wavelengths below 700nm. Figure 12 shows the integrated intensity in the image plane of a confocal microscope for a dipolar scatterer as a function of defocus [25]. Here the x axis is $4kz\sin(\alpha/2)^2$ where z is the defocus and $\sin(\alpha/2)=NA$. One can see that if the intensity in the image plane has decreased to exp(-1) the approximate expression

$$5 = 8\pi \frac{z_{1/e}}{\lambda} NA^2 \rightarrow z_{1/e} \approx 0.2 \frac{\lambda}{NA^2} \qquad (30)$$

holds. For a wavelength of 800nm and a NA of 1.4 this leads to a $z_{1/e}$ of only 80nm. In this calculation the annular illumination and the presence of the glass/air interface is not accounted for and are expected to even decrease this value. From this calculation it is expected that deviations from the ideal focal position in the order of 50nm already lead to a significant drop in intensity in the image plane. Clearly any chromatic aberration with the same order of magnitude will lead to an additional transmission function $T(\lambda,z)$ which is superimposed on the actual spectrum.

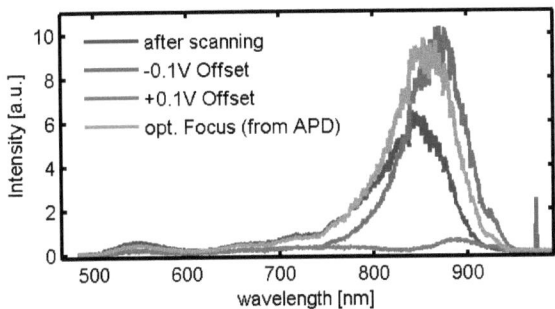

**Figure 11: Typical scattering spectra from crescent shaped resonators as function of z-position (the control voltage is 0.125V/μm)**

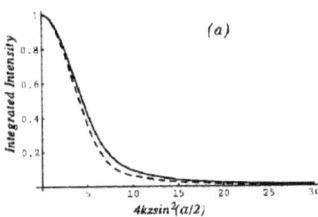

**Figure 12: Integrated Intensity as a function of defocus[25]**

To find out the chromatic aberration present in the microscope a scan of the focal position against a mirror was carried out and the reflected signal was recorded with the spectrometer for different focal positions. Then the spectra where normalized by the spectrum which gives the highest integrated intensity (z = 0) and the position of maximum intensity per wavelengths was extracted. The result is shown in figure 13. It can be seen that below 700nm almost linear achromatic aberration with a slope df/dλ=−1μm/100nm is present. This explains the irreproducibility of the spectra when tuning the focus on the integrated intensity. Depending on the spectral position of a resonance peak a different focus will be adjusted, and this leads to an additional transmission function which cuts the spectrum at the low wavelength side, which leads to apparent shifts in the spectral position of the resonance.

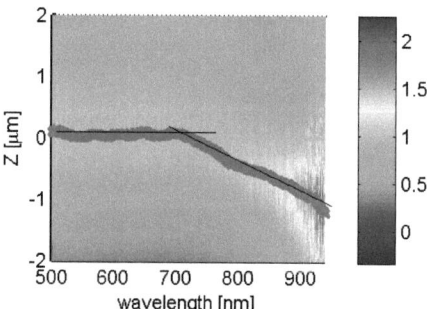

**Figure 13: measured relative transmission of the confocal microscope (relative to t=0). The red line is the line of maximum transmission as a function of wavelength**

The solution to this problem is to adjust the focus only with a very narrowband part of the total spectral signal, which should be ideally centered at the resonance position. For this reason a bandpass filter was added in front of the APD and scanning and adjustment of the focal position was done only with this filter. The used filter has a centre wavelength of 800nm and a width of 10nm. The loss of signal intensity turned out to be tolerable (~80% loss). The filter has the additional property to block signals from resonators which are not close to 800nm with their resonance. That means that only a subgroup of physically similar resonators will be visible on a scanned image.

With this approach the spectral reproducibility increased to less than 1nm. Clearly it would be desirable to have achromatic optics in the NIR, however no chromatically corrected high NA objectives in the NIR exist currently. One solution would be to use a mirror objective which naturally has zero chromatic aberration. A drawback of this approach would be the loss of NA as mirror objectives are only available up to 0.6 NA. Also the other part of the microscope like the confocal pinhole would have to be achromatized too over a wide bandwidth.

### 3.1.3.2. Finding single scatterers

Another practical problem was to find single resonators in the SEM or AFM after optical measurements were carried out. This is not hard to imagine since usually the sample sizes are in the order of square cm but the optically observed areas is only some ten microns wide. The first approach to solve this problem was to simply make a small scratch onto the sample and use the resonators in the vicinity of this scratch. This works well when there are only very few resonators on an otherwise clean sample. When there are more objects on the sample like additional colloids the problem is that usually the images in SEM have a low contrast and it is hard to really distinguish positions on the sample unambiguously (figure 14).

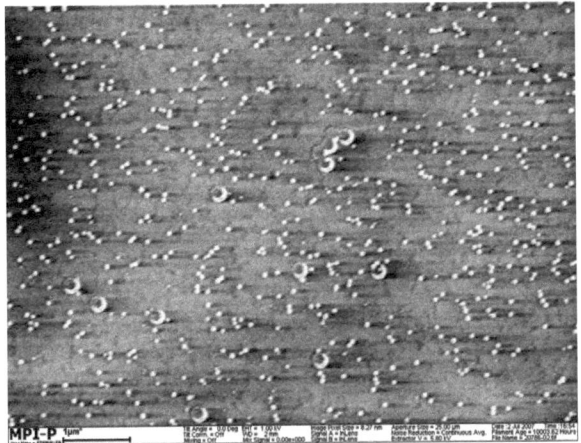

**Figure 14: Typical SEM image of crescents with colloids.**

What is really desirable is to have some kind of coordinate system structured onto the sample. The approach used here was to mount a TEM grid onto the sample after fabrication of the resonators and evaporate gold trough it. This results in additional gold pads with free lines between them. Imaging was done between these pads. Additionally a scratch was also made into the pads to give coarse orientation.

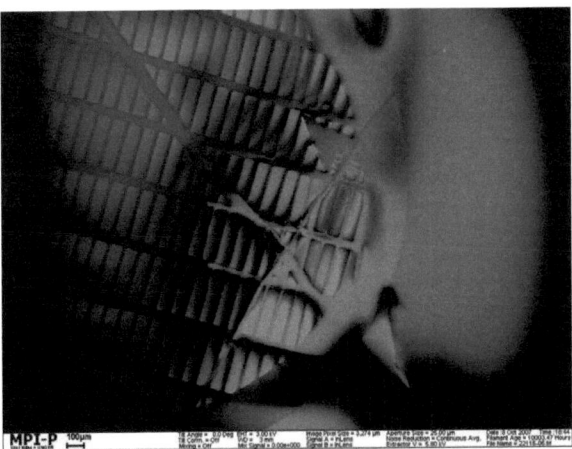

**Figure 15: SEM image of the evaporated TEM Grid and an additional scratch used for finding positions on the sample unambiguously.**

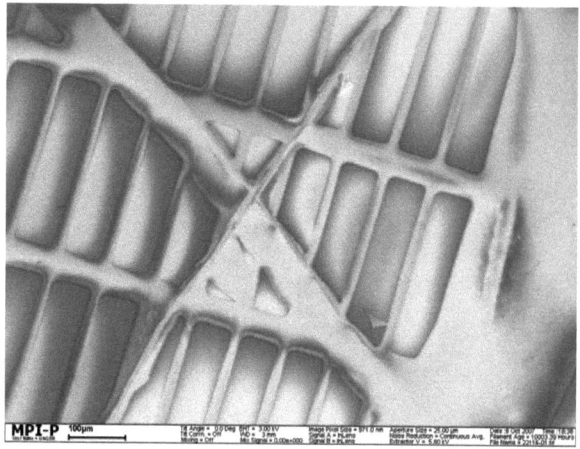

**Figure 16: Close up of the grid. Scanning is done between two gold pads close to the scratch.**

Figure 15 and 16 show such an evaporated and scratched TEM grid on a sample. The mesh size of the TEM grid was 300x75μm with 50μm spacing between the pads. This is the optimum distance because now the imaging can be done between two pads and the maximum scanning distance of the piezo table of 80μm is optimally used.

## 3.2. FDTD

For the most numerical calculations presented in this work the finite difference time domain method (FDTD) is used. Here Maxwell's curl equations are approximated by finite difference equations on a regular grid in space. The electric and magnetic field are evolved in time using an explicit time marching scheme. The method is described extensively by Taflove [26] and has been used earlier to calculate the electromagnetic properties of metallic nanoparticles [27]. The equations that are discretized are the two Maxwell's equations

$$\mu_0 \mu_R \dot{\vec{H}} = -\nabla \times \vec{E} - \vec{M}$$
$$\varepsilon_0 \varepsilon_R \dot{\vec{E}} = \nabla \times \vec{H} - \vec{J} - \dot{\vec{P}}$$
(31)

Here H is the magnetic field, E is the electric field, M is a magnetic current density, H is the magnetic field, J is an electric current density and P is an electric polarization. In the second equation usually the time derivative of D is on the left side. This was divided here in a D which results from a constant but possibly space dependent permittivity and a polarization term P which reflects the contribution of dispersive permittivities.

$$\vec{D}(t) = \varepsilon_0 \varepsilon_R \vec{E}(t) + \vec{P}(t) = \varepsilon_0 \varepsilon_R \vec{E}(t) + \varepsilon_0 (\varepsilon(t) - \varepsilon_R) \vec{E}(t)$$
$$\vec{B}(t) = \mu_0 \mu_R \vec{H}(t) + \vec{M}(t) = \mu_0 \mu_R \vec{H}(t) + \mu_0 (\mu(t) - \mu_R) \vec{H}(t)$$
(32)

This way of splitting the material contributions reflects better the way materials are actually implemented in FDTD methods. Direct treatment of dispersive material functions in FDTD would lead to a convolution which is computationally expensive. Instead of doing this convolution directly it is computationally less demanding to solve the governing differential equation for the impulse response of P and M simultaneously with equations (31) (see for example [26]). The current densities M and J consist of free currents and therefore source of the fields $J_{source}$ and $M_{source}$ and a term which reflect conductive materials

$$\vec{J} = \sigma \vec{E} + \vec{J}_{Source}$$
$$\vec{M} = \sigma_M H + \vec{M}_{Source}$$
(33)

There are many possibilities of discretizing these equations. The most common one is known as the Yee Algorithm [28]. This algorithm uses two different regular grids for E and H, which are shifted half a step and are evaluated also shifted in time about half a step.

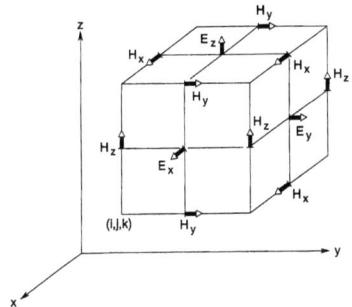

**Figure 17: The space lattice in the Yee algorithm on which E and H are evaluated [28].**

The grid used in the Yee algorithm is shown in figure 17. It can be interpreted as a discretized version of Faradays and Amperes law. A discretized quantity in time and space is noted as

$$u(i\Delta x, j\Delta y, k\Delta z, n\Delta t) = u_{i,j,k}^n$$
(34)

Equations (31) are discretized using centered differences in space and time leading to

$$E_x\Big|_{i,j+1/2,k+1/2}^{n+1/2} = c_1 E_x\Big|_{i,j+1/2,k+1/2}^{n-1/2} +$$

$$c_2 \left( \frac{H_z\Big|_{i,j+1,k+1/2}^{n} - H_z\Big|_{i,j,k+1/2}^{n}}{\Delta y} - \frac{H_y\Big|_{i,j+1/2,k+1}^{n} - H_y\Big|_{i,j+1/2,k}^{n}}{\Delta z} - J_{Source,x}\Big|_{i,j+1/2,k+1/2}^{n} \right)$$

with

$$c_1 = \left( \frac{1 - \frac{\sigma_{i,j+1/2,k+1/2} \Delta t}{2\varepsilon_{i,j+1/2,k+1/2}}}{1 + \frac{\sigma_{i,j+1/2,k+1/2} \Delta t}{2\varepsilon_{i,j+1/2,k+1/2}}} \right) \tag{35}$$

$$c_2 = \left( \frac{\frac{\Delta t}{\varepsilon_{i,j+1/2,k+1/2}}}{1 + \frac{\sigma_{i,j+1/2,k+1/2} \Delta t}{2\varepsilon_{i,j+1/2,k+1/2}}} \right)$$

for $E_x$ and

$$H_x\Big|_{i-1/2,j+1,k+1}^{n+1} = c_3 H_x\Big|_{i-1/2,j+1,k+1}^{n} +$$

$$c_4 \left( \frac{E_y\Big|_{i-1/2,j+1,k+3/2}^{n+1/2} - E_y\Big|_{i-1/2,j+1,k+1/2}^{n+1/2}}{\Delta z} - \frac{E_z\Big|_{i-1/2,j+3/2,k+1}^{n+1/2} - E_z\Big|_{i-1/2,j+1/2,k+1}^{n+1/2}}{\Delta y} - M_{Source,x}\Big|_{i-1/2,j+1,k+1}^{n+1/2} \right)$$

with

$$c_3 = \left( \frac{1 - \frac{\sigma^*_{i-1/2,j+1,k+1} \Delta t}{2\mu_{i-1/2,j+1,k+1}}}{1 + \frac{\sigma^*_{i-1/2,j+1,k+1} \Delta t}{2\mu_{i-1/2,j+1,k+1}}} \right) \tag{36}$$

$$c_4 = \left( \frac{\frac{\Delta t}{\mu_{i-1/2,j+1,k+1}}}{1 + \frac{\sigma^*_{i-1/2,j+1,k+1} \Delta t}{2\mu_{i-1/2,j+1,k+1}}} \right)$$

for $H_x$. The Other Field components are analogous (and simply found by cyclic permutation) but are omitted here to save space. From these equations it can be seen that E is evaluated at half time steps while H is evaluated at integer time steps. On each point on the space grid only one field component is stored.

FDTD is, beside errors that arise from the discretisation, an exact method, which can solve in principle any wave propagation problem to any desired accuracy but is limited by the available computational resources. For the simulation of metallic nanoparticles this limitation is usually severe because of the different length scales involved. The electric field in a metal particle has a typical penetration depth of some nm while the particle is often 100nm in size needing computational cells of double the size plus an additional space for the PML to eliminate reflections at the cell boundary. This adds up to computational cells of around one micron in size which have to be discretized with nm grid size. In 3D this leads to around one billion grid points where six field components plus material properties have to be stored. This leads to memory requirements above 10GB RAM. Even more severe is that a stability criterion limits the time step of the simulation. The maximum allowable time step is proportional to the spatial grid size, a fact which is known as Courant condition of stability. As a result for a given size of the computational cell the computation time scales with $\Delta x^4$ which leads for fine discretisations to computation times in the order of days. The Yee algorithm is second order accurate due to its centered difference scheme.

For the calculations done in this work the freely available code meep [29] was applied. This code provides the basic time stepping algorithm, boundary conditions (PML), an implementation of dispersive materials including Drude-Lorentz materials and an implementation of electric and magnetic current sources. A desirable implementation of a plane wave in form of a total field/scattered field (TFSF [26]) boundary is missing and was therefore implemented. The following paragraph describes this implementation.

A TFSF boundary divides the simulated region in two areas: one with the total field and one where only the scattered field is present (figure 18). On the boundary between these two areas the analytical known incident field is subtracted from the total field outside of the total field area.

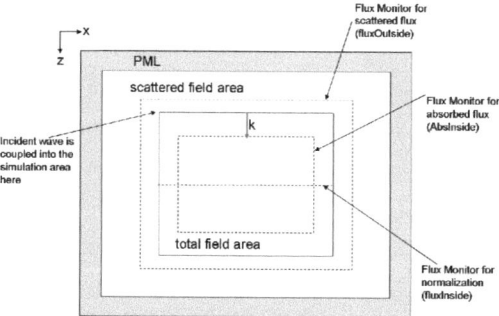

**Figure 18: Construction of the computational cell with TFSF boundary and Monitors for the integrated total and scattered flux. In brackets are the variable names used in the actual code which is shown in Appendix B.**

$$(36)$$

This subtraction leads to an additional term in the time stepping equations (35) and . This terms are given [26] as (again only for the $E_x$ and $H_y$ equations at a y-boundary):

$$E_x\Big|_{i,j0,k}^{n+1} = \left\{E_x\Big|_{i,j0,k}^{n+1}\right\}_{Eq.32} - \frac{\Delta t}{\varepsilon_0 \Delta} H_{z,inc}\Big|_{i,j0-1/2,k}^{n+1/2}$$

$$H_x\Big|_{i,j0-1/2,k}^{n+1/2} = \left\{H_x\Big|_{i,j0-1/2,k}^{n+1/2}\right\}_{Eq.33} - \frac{\Delta t}{\mu_0 \Delta} E_{z,inc}\Big|_{i,j0,k}^{n}$$

$$(37)$$

Here $H_{inc}$ and $E_{inc}$ are the incoming fields which are usually a plane wave. By comparing (35) and (36)

and with (37) one can see that the additional terms of the incoming waves are equivalent to a source term, either a magnetic or an electric current. For example $H_{z,inc}$ corresponds to an electric current source $J_x$.

$$\frac{\Delta t}{\varepsilon_0 \Delta} H_{z,inc} = \left( \frac{\frac{\Delta t}{\varepsilon_{i,j+1/2,k+1/2}}}{1 + \frac{\sigma_{i,j+1/2,k+1/2} \Delta t}{2\varepsilon_{i,j+1/2,k+1/2}}} \right) J_{Source,x} \qquad (38)$$

This correspondence gives a very simple way to implement a TFSF source when electric and magnetic current sources are available as sources. One just has to place for each point on the TFSF boundary the corresponding current term, which is given by equations similar to (37) from literature [26]. In this way the TFSF source was implemented in this work. The used equations are not shown here to save space. The resulting code for a plane wave travelling in z direction is shown in Appendix B. In figure 18 the variable names used to calculate the scattered and absorbed powers are also shown.

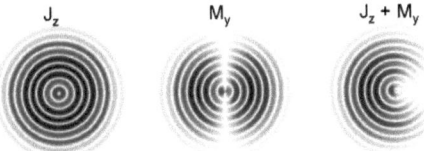

**Figure 19: Superposition of electric and magnetic currents.**

The equivalence of a current sources and TFSF source can also be understood from a physical point of view. Figure 19 shows the electric field pattern in the x-y plane of an electric and a magnetic current point source. These have different symmetries and hence a superposition leads to interference and the cancelling of the radial wave in one direction and constructive interference in the other. By placing many of this sources in a line a plane wave is constructed. Because of the different symmetries of electric and magnetic currents it would be possible to construct any field distribution regardless of the diffraction limit. This is also a strong motivation to develop magnetic materials in the visible, so called metamaterials.

The following paragraph describes the setup of a simulation in meep. The simulation area is surrounded by a perfectly matched layer (PML, [30]) which absorbs the outgoing waves and for excitation a total field/scattered field boundary was used. Figure 20 shows a cut through the simulated space with source, absorbing boundary and the particle. At the TFSF boundary (red cuboid around the particle) currents are distributed in a way that without the particle inside the cuboid (volume 2) a x-polarized Gaussian light pulse with a center wavelength of 900 nm and a wavelength-bandwidth of 2400 nm propagates in positive z direction and is partially reflected by the glass surface. Without the scattering particle, no field is present outside the cuboid (volume 1).

The presence of a particle changes both the field inside and outside the TFSF boundary. Integration of the flux through a closed surface outside the TFSF boundary (dashed line) yields the total scattered flux and the scattering cross section. Absorbed flux and absorption cross section are obtained in an analogous way by integration over a closed surface inside the TFSF boundary (dotted line). The size of the computational area as well as the resolution was chosen such that convergence of the results was observed, indicating their independence of the mentioned parameters. A small residual error in the order of 5% remains due to reflections at a non-perfect PML and manifests itself as an overlaying oscillation of the cross sections especially at smaller wavelengths. Because this effect is small and is the same in all simulations with the same box/PML size it has only little influence on finding the resonances from the simulation. The computational box used here has the size 0.8x0.8x0.7 µm, the PML is 100 nm thick around the whole box and the TFSF boundary is of 0.5x0.5x0.4 µm size.

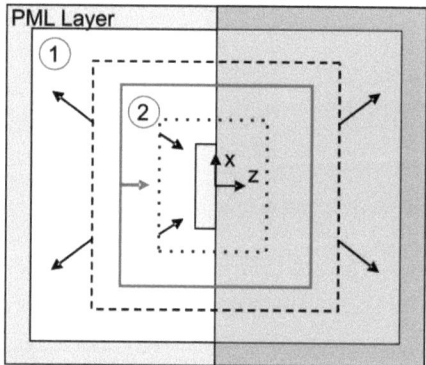

**Figure 20: 2D cut along the x-z plane of the three-dimensional simulation space. The yellow object symbolizes the particle which lies on a glass surface. The TFSF-boundary is sketched as red line, surfaces for integration as black dashed and dotted lines.**

The discretisation of sharp corners deserves special attention. Theoretically, at an infinitely sharp corner the field diverges. From the fabrication process corner radii of no smaller than 5 nm are assumed. It is known that for gold nanoparticles which are smaller than this radius, the effective permittivity deviates from the bulk permittivity of gold due to several physical mechanisms summarized as finite size effects [21, 22]. They provide an approximate lower limit for feature sizes where the metal can be described by its bulk response. Figure 21 shows possible discretisations for a tip for a pixel size of 5 nm. The tip was always constructed as shown on the left where an effective radius of 5 nm is approximately reproduced. Single pixels with metallic response as shown on the right were avoided by manual reshaping of the tip after the geometric construction. Actually it still is questionable that the field values directly at the tip are correctly calculated, because still the tip is only described by very few grid points. It would be really desirable to be able to use a locally refined mesh in the FDTD algorithm. Comparisons with FEM calculations later showed however, that some nm away from the tip the field values are quite accurate and deviate from the FEM results less than 10 percent.

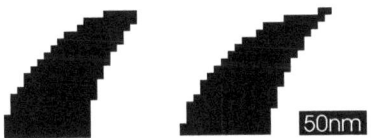

**Figure 21: close-up of two discretized tips. Each pixel corresponds to a simulation cell (white corresponds to the permittivity of air, black to that of gold). The left configuration is used in all simulations presented here**

For modelling the dielectric properties of gold, a two oscillator Drude-Lorentz model was used. The first oscillator accounts for the contribution of the free electrons, while the second describes an inter-band transition of the gold. The formula for the dielectric function is:

$$\varepsilon = \varepsilon_\infty + \frac{\Delta\varepsilon_1}{-\omega^2 - i\Gamma_1\omega} + \frac{\omega_2^2 \Delta\varepsilon_2}{\omega_2^2 - \omega^2 - i\Gamma_2\omega} \qquad (39)$$

The values used are: $\varepsilon_\infty = 9.17$, $\Delta\varepsilon_1 = 52.26$, $\Gamma_1 = 0.057$, $\omega_2 = 1.65$, $\Delta\varepsilon_2 = 0.068$, $\Gamma_1 = 0.062$. All dimensions are $2\pi c/\mu m$. The dielectric response as obtained with this model in comparison to tabulated data is shown in figure 22. For modelling the glass support, a constant permittivity of 2.25 is used

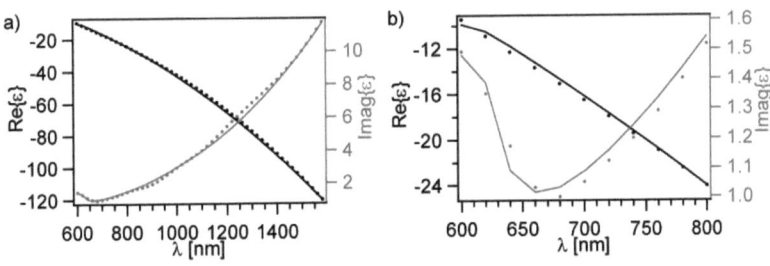

**Figure 22: Dielectric function of gold from [20] (dots) and fitted dielectric function (continuous lines) The values were obtained by fitting the data from [20] with a standard nonlinear least squares routine.**

## 3.3. FEM

A second method used for simulations in this work is the finite element method (FEM). The name finite element method comes from the fact that in this method the solution of a partial differential equation (PDE) is approximated with a sum of functions that are only defined on small elements or patches in space, which are for example triangles in 2D or tetrahedra in 3D. For the original PDE a variational formulation is derived and an ansatz consisting of these functions is inserted. The result is a large but sparse system of linear equations which has to be inverted to find a solution. FEM is usually used in the frequency domain. For the FEM method there exists a large literature base, see for example [31] for a derivation of the method used here. The FEM method has for plasmonic resonators several advantages over FDTD: It is possible to use unstructured grids to discretize space, which means that small features can be discretized very fine to resolve strong field gradients there but other space where no strong field gradients are expected can be discretized much coarser. This is ideal for the multiscale nature of plasmonic resonators and saves computation time and memory. Due to the solution of the PDE in the frequency domain, there are no stability criteria like in the time domain. Because the solution of the linear system of equations scales almost linear in the number of unknowns, for a fixed computational domain the algorithm scales with $\Delta x^3$ which is one order better then FDTD. Practically the algorithm scales even better, because usually not the whole domain has to be refined homogeneously. In this work the software package JCMwave was used [31]. This software can calculate scattering or eigenvalue problems in 2D, 2D with radial symmetry and 3D domains. It has already open boundaries (PMLs) included and can do adaptive mesh refinement based on local error estimation. The software is used via a set of matlab scripts which construct the geometry, call the solver and do the postprocessing. For postprocessing various possibilities are already included in the software, e.g. calculation the H field from the E field or calculating the far field after propagating trough an optical system.

# 4. Sensing theory

In this chapter the sensing theory will be developed. The key question here is: How can the process of detecting a local change in refractive index be described? All optical label free sensing techniques will translate such a change into a measured signal which can be for example the shift of a spectral peak or the change of the intensity on a detector. Because of this general principle, a theory describing the response of a nanoplasmonic transducer and finally the detectability of a signal is very general and not even limited to this specific case. This means that the generality of the theory enables also the comparison of different methods. The chapter is structured in the following way: First a figure of merit (FOM) is defined, which reflects directly the ability of a sensor to detect a given event. Then, based on this FOM, the effect of noise is investigated. Then an analytical expression for the sensitivity of a resonator is found and verified. The application and limitations of this expression is demonstrated with a simple example. A specialization of this expression to the quasi-static case is shown next. Then the FOM is expressed in the found analytical expressions which gives the connection between the resonator properties and the FOM. The found FOM is compared for different sensing concepts. Last the scaling of this FOM with size and shape of a quasi-static resonator is discussed in detail. Partly the results of this chapter have been published already in scientific journals. The general derivation of the FOM and it's scaling properties have been published in [32] and the results of this chapter have been used to interpret experimental results in [33, 34].

## 4.1. Derivation of a Figure of Merit for the sensitivity

For a given sensing event which is to be detected there will be a change in a detector signal A which is called the response $\Delta A$. The response will be a function of the transducer and the detected event, for example the binding of a molecule to the surface of a plasmonic resonator. Because of unavoidable imperfections and physical limitations the measurement of $\Delta A$ cannot be exact but will suffer from noise. The noise will introduce an uncertainty in the detection $\Delta A_{Un}$. A general Figure of merit of the sensing process can now be how much better than this uncertainty the event can be detected. This leads to the FOM

$$FOM \propto \frac{\Delta A}{\Delta A_{Un}} \qquad (40)$$

This FOM can also be regarded as the final signal to noise ratio of the measurement. Because the usual abbreviation SNR will be used for other signal to noise ratios in this work, this ratio will only be called FOM.

In the following this term is applied to the case of plasmonic resonators and a specific detection scheme. Because the resonance is detected as a distinct peak in scattering or absorption spectra the measured signal is often the center frequency $\omega_C$ of a spectral peak and the response is a change $\Delta \omega$ which is measured with an uncertainty $\Delta \omega_{un}$.

$$FOM \propto \frac{\Delta \omega}{\Delta \omega_{Un}} \qquad (41)$$

To find what properties of the resonator affect the FOM analytical expressions for $\Delta\omega$ and $\Delta\omega_{Un}$ have to be found. First the uncertainty will be considered.

## 4.2. Measurement uncertainty

It is assumed that a spectral measurement of the intensity as a function of frequency is done. The center frequency $\omega_C$ is then found by fitting that measurement to the known line shape F of the resonance. In many cases the measurement will be carried out with a grating spectrometer coupled to a CCD so the signal will be the number of counts $S_n$ for a discrete set of frequencies $\omega_n$. Each of these intensity measurements will have a measurement error $\sigma_n$. A standard way of fitting this dataset to the known line shape is minimization of the weighted sum of squares

$$\chi^2 = \sum_{n=1} \left( \frac{S_n - SF(\omega_n, \Gamma, \omega_C)}{\sigma_n} \right)^2 \tag{42}$$

where the weighting is done with the inverse of the actual error to ensure that highly accurately measured points will contribute more to the result than inaccurate ones. S is the peak signal and $\Gamma$ the full width of the peak at half the peak signal (FWHM). The minimum of $\chi^2$ is found by varying the parameters $\omega_C$, $\Gamma$ and S in a systematic way for example with a mathematical algorithm. To find $\Delta\omega_{un}$ the question is then how $\omega_C$ is affected by the measurement errors $\sigma_n$. This is actually an old question and has been answered for example by Bobroff [35]. To understand the results later a short outline of his reasoning is given here.

After finding the minimum $\chi^2_{min}$, the sum of squares $\chi^2$ is linearized in the parameters around $\chi^2_{min}$ and the difference $\Delta = \chi^2 - \chi^2_{min}$ is formed. $\Delta$ is a measure of the likelihood that a set of parameters $\omega_C$, $\Gamma$ and S, which is differing from $\omega_{CMin}$, $\Gamma_{Min}$ and $S_{Min}$, is the true result of the measurement. For a given confidence level of the parameters $\Delta$ can be calculated [36]. The result is

$$\Delta = \sum_{n=1} \frac{1}{\sigma_i^2} \left( \frac{\partial SF(\omega_n, \Gamma, \omega)}{\partial \omega} \bigg|_{\omega_c} \right)^2 \Delta\omega_{un}^2 \tag{43}$$

This is now simplified by dividing by the number of channels of the detector N and expressing 1/N by the frequency spacing between the channels $\delta\omega$ and the total frequency range of the detector $\omega_{Det}$: $1/N = \delta\omega/\omega_{Det}$. Inserting this into (43) yields

$$\frac{\Delta}{N} = \frac{\delta\omega}{\omega_{Det}} \sum_{n=1} \frac{1}{\sigma_i^2} \left( \frac{\partial SF(\omega_n, \Gamma, \omega)}{\partial \omega} \bigg|_{\omega_c} \right)^2 \Delta\omega_{un}^2$$

$$\approx \frac{\Delta\omega_{un}^2}{\omega_{Det}} \int_{-\omega_{Det}/2}^{\omega_{Det}/2} \frac{1}{\sigma(\tilde{\omega})^2} \left( \frac{\partial SF(\tilde{\omega}, \Gamma, \omega)}{\partial \omega} \bigg|_{\omega_c} \right)^2 d\tilde{\omega}$$

(44)

here the sum can be approximated with an integral, when the channel spacing is narrow. This seems to be a complication first but for some line shapes and noise cases this integral can be solved analytically and this leads to significant simplifications and insight into the problem. An important fact can be deduced already from (43) or (44): The error of each point is weighted by the derivative of F with respect to ω which means that the positions where F is steepest contribute most to the accuracy of the measurement.

Having (44) at hand makes it possible to determine $\Delta\omega_{Un}$ for a given confidence level, experimental setup, line width and error distribution. Next the noise of the measurement is considered. Different noise sources can be distinguished. First an unavoidable noise source is the counting statistics of the counting device and the noise statistics of the light itself. Events occur independently at a constant average rate at the detector. This leads to a Poissonian distribution which has the standard deviation and therefore a given noise of $S^{1/2}$ when S is the number of counted photons. The SNR of this noise case is $S^{1/2}$ as well. This noise case will be called signal noise. For a Gaussian line shape the evaluation of (44) is given as [35]:

$$\Delta\omega_{un} = \Delta\omega_{SN} = \frac{\Gamma}{\sqrt{S}} \sqrt{\frac{\Delta}{N_\Gamma tF_p}}$$

(45)

where $N_\Gamma$ is the number of detector channels per FWHM and $tF_p$ is given as

$$tF_p = \int_{-t}^{t} u^2 e^{-u^2} du$$

(46)

and t is the dimensionless frequency range of the detector $t=\omega_{Det}/2\Gamma$. The case of a Lorentzian line shape which is relevant for this work is derived in Appendix A but is the same except for the integral $tF_p$. From (45) it can be seen nicely which parameters actually determine the uncertainty $\Delta\omega_{un}$. From the properties of the resonator it is the line width of the resonance and the SNR $S^{1/2}$ which is connected to the resonator via its cross section $\sigma_P$. Depending on the experiment $\sigma_P$ can be a scattering cross section $\sigma_{Scat}$ or an extinction cross section $\sigma_{Ext}$. For the instrument the incoming intensity I, the transmittance of the Instrument T, the detection efficiency of the detector P and the integration time t have to be considered, and the Signal S is given by

$$S = P \cdot T \cdot I \cdot \sigma_p \cdot t \tag{47}$$

Of further importance is the number of detector channels $N_\Gamma$ and the dimensionless width of the resonance line $tF_p$ which is used for fitting. An interesting property of (45) is that it is not possible to change the uncertainty by just changing the number of detection channels. When the number of detector channels is doubled then the signal S in each channel will be only half. Since $\Delta\omega_{un}$ has the same dependence on S and $N_\Gamma$ it will be unaffected. This is not necessarily so for other noise cases. For example when the dominating noise level is constant and independent of signal intensity it makes sense to use fewer channels which collect more light.

Equation (45) suggests a general form of the uncertainty

$$\frac{\Delta\omega_{un}}{\Gamma} = \frac{f(\Delta,t)}{SNR} \tag{48}$$

for other cases of noise. For some other cases this can also be shown explicitly. For example in the case of a constant additive noise as the dominating noise source (which will be called background noise) $\sigma_B$ it is given by [35]

$$\Delta\omega_{un} = \Delta\omega_{BN} = \frac{\Gamma}{S/\sigma_B} \sqrt{\frac{\Delta}{N_\Gamma tF_p}} \tag{49}$$

Another important case of noise is an apparent shift of the center frequency $\omega_C$. Such errors, which will be called instrument noise, can be caused for example by thermal fluctuations of the environment of the resonator or fluctuations in the focal position of a microscope which is not well corrected for chromatic aberrations (shown in chapter 3.1.3.1). Obviously such errors are independent of the resonator, and if a measurement under such conditions has to be carried out the response $\Delta\omega$ alone has to be maximized. For a generalized treatment of all noise cases these errors can be translated to intensity noise by

$$\sigma_i = S \left. \frac{\partial F(\omega_n, \Gamma, \omega)}{\partial \omega} \right|_{\omega_C} \Delta\omega_C \tag{50}$$

where $\Delta\omega_C$ is he fluctuation of the center frequency.

## 4.3. An analytical expression for the peak shift

### 4.3.1. Weakly radiating systems

Having an expression for the measurement uncertainty the next step is to find a relation between resonator properties like the near field enhancement and the quality factor and the response to an analyte. The electric field of the resonator is a solution to the vector wave equation

$$L\vec{E} + \frac{\varepsilon}{c^2}\ddot{\vec{E}} = 0 \text{ with } L\vec{E} = \nabla \times (\nabla \times \vec{E}) \quad (51)$$

Solutions to (51) are of the form

$$\vec{E} = \vec{E}_0(\vec{r})e^{-i\omega t} \quad (52)$$

These solutions describe time harmonic waves with an angular frequency $\omega$.
From first order perturbation theory it is known (see for example [37]), that in the case of a small permittivity change $\Delta\varepsilon(r)$ a new solution with a change $\Delta\omega$ in $\omega$ can be found:

$$\frac{\Delta\omega}{\omega} = -\frac{1}{2}\frac{\int_{V_s}\Delta\varepsilon\vec{E}^2 dV}{\int_{AllSpace}\varepsilon\vec{E}^2 dV} \quad (53)$$

This is easiest shown with the help of the variational formulation of the wave equation (51). In the case of a lossy nonradiating nondegenerate eigenvalue this reads as [37]:

$$\omega^2 = \frac{\int \vec{E}L\vec{E}dV}{\int \vec{E}\varepsilon\vec{E}dV} \quad (54)$$

because this is a variational expression in E and hence stationary, small changes in E will not change $\omega$ in first order. It is important to note that to derive (54) the usual inner product with complex conjugation cannot be used because the resonators considered here are usually lossy and the problem is therefore not Hermitian. This leads to eigenfrequencies which are in general complex. To arrive at (53) one only has to insert the perturbed permittivity as $\varepsilon = \varepsilon_0 + \eta\Delta\varepsilon$ and differentiate for $\eta$. The result which is also known as Hellmann-Feynman theorem (see for example [38]) is exact in the limit of infinitesimal small $\Delta\varepsilon$ which means that the slope $\partial\omega/\partial\varepsilon$ is exact. Since changes in the permittivity for usual sensing applications are small, deviations from this approximation can usually be neglected. Lai et al. also showed that the perturbation expression is still valid when the permittivity change is high but the integral over $\Delta\varepsilon E^2$ is small [39].

## 4.3.2. Strongly radiating systems

Problems with (53) arise with strongly radiating systems. When radiation cannot be neglected there exist spherical waves in the far field of the resonator with radial dependence exp(ikr)/r where r is the radial coordinate of a spherical coordinate system. The wave vector k is usually complex and the condition that it describes outgoing which are decaying in time waves leads to Im(k)<0. This results in a solution which grows towards infinity and hence the denominator does not converge and (53) becomes zero. In order to overcome this problem the radiation field has to be considered in the perturbation theory. This was done by Lai et. al. [39] who also included the case of degenerate modes. The result is an additional surface integral in the denominator of (53) over the boundary of the considered space (for the nondegenerate case):

$$\frac{\Delta\omega}{\omega} = -\frac{1}{2} \frac{\int_{V_S} \Delta\varepsilon \vec{E}^2 dV}{\int_{AllSpace} \varepsilon \vec{E}^2 dV + \frac{ic}{2\omega}\int_{Boundary} \vec{E}^2 dA} \tag{55}$$

They showed that this additional term exactly cancels out the outgoing fields and gives the proper perturbation expression when the integral is carried out in the far field. As the surface term is proportional to the energy of a radiated spherical wave, this can be interpreted in a way that the radiated energy does not contribute to any frequency change. This is an intuitive result because once this energy is radiated away it cannot interact with the perturbation anymore. Equation (55) suggests therefore to neglect the contribution of the radiating parts of the field. That means that the volume integral should only be carried out where the near field is dominating and the rest can be neglected. For plasmonic resonances this is easy because of the high field enhancement in the near field. As the surface term is proportional to the radiated energy and, as will be shown later, the volume term proportional to the stored energy, the ratio of them will be proportional to the quality factor Q when Q is dominated by radiation and not absorption. Hence neglecting the surface term will lead to errors in the order of 1/Q. Since the quality factors of good plasmonic resonators are usually better than 10, these errors will be smaller than 10%. Numerical experiments will later show that they are usually smaller. From practical considerations systems with strong radiation and low Q are not interesting for sensing applications, which will be shown later. Therefore for all practical applications the use of (53) instead of (55) is valid.

## 4.3.3. Expression in terms of refractive index and interpretation

An interpretation of (53) was shown in [40]. When $\Delta\varepsilon$ is approximated by $\Delta\varepsilon \approx \varepsilon 2\Delta n/n$ where $\Delta n$ is the refractive index change, n is the refractive index and the perturbation is assumed constant over the perturbed volume:

$$\frac{\Delta\omega}{\omega} \approx -\frac{\Delta n}{n}\frac{\int_{V_s}\varepsilon\vec{E}^2 dV}{\int_{AllSpace}\varepsilon\vec{E}^2 dV} = -\frac{\Delta n}{n}C \qquad (56)$$

For transparent materials with negligible dispersion the term in the integral is the energy density of the field so this shows that the relative change in frequency is proportional to the relative change in refractive index and the fraction of the energy C within the detection volume where the change takes place. This fraction will be called energy confinement.

### 4.3.4. Generalization for strongly dispersive resonators

For strongly dispersive resonators equation (56) does not represent an accurate result. Here the frequency dependence of the permittivity of the resonator leads to an additional change in resonance frequency. In particular for quasi-static resonances, the resonance condition leads directly to a condition for the resonators permittivity (see chapter 2.1.2), which shows that it is impossible to obtain a resonance without dispersion. This case can be covered by perturbation theory as well. To do so, a new permittivity perturbation which accounts for the change of permittivity of the resonator

$$\Delta\tilde{\varepsilon} = \Delta\varepsilon + \frac{\partial\varepsilon}{\partial\omega}\Delta\omega \qquad (57)$$

is inserted into (53). With some algebraic manipulation (53) becomes

$$\frac{\Delta\tilde{\omega}}{\omega} = -\frac{1}{2}\frac{\int_{V_s}\Delta\varepsilon\vec{E}^2 dV}{\int_{AllSpace}\varepsilon\vec{E}^2 dV + \frac{1}{2}\int_{AllSpace}\omega\frac{\partial\varepsilon}{\partial\omega}\vec{E}^2 dV} \qquad (58)$$

Now one can replace $\frac{1}{2}\omega\partial\varepsilon/\partial\omega = \frac{1}{2}\partial(\omega\varepsilon)/\partial\omega - \frac{1}{2}\varepsilon$ and arrives at

$$\frac{\Delta\tilde{\omega}}{\omega} = -\frac{\int_{V_s}\Delta\varepsilon\vec{E}^2 dV}{\int_{AllSpace}\varepsilon\vec{E}^2 dV + \int_{AllSpace}\frac{\partial\omega\varepsilon}{\partial\omega}\vec{E}^2 dV} \qquad (59)$$

which can be again cast to the form of (56):

$$\frac{\Delta\tilde{\omega}}{\omega} = -2\frac{\Delta n}{n}\frac{\int_{V_s}\varepsilon\vec{E}^2 dV}{\int_{AllSpace}\varepsilon\vec{E}^2 dV + \int_{AllSpace}\frac{\partial\omega\varepsilon}{\partial\omega}\vec{E}^2 dV} = -2\frac{\Delta n}{n}\frac{V}{G_1 + G_2} = -2\frac{\Delta n}{n}\tilde{C} \qquad (60)$$

This expression which includes dispersion effects of the resonator differs from (56) in the vanishing of the prefactor ½ and the appearance of a new term $G_2$ which is the expression of the energy of the electric field in an dispersive medium far away from a resonance [41]. Without dispersion, (56) and (60) are identical.

### 4.3.5. Discussion of the quasi-static case

The implications of (60) will now be discussed. To simplify the discussion the quasi-static case will be considered first. In this case the resonator is much smaller than the wavelength of light so that retardation effects are negligible. In this case Maxwell's equations for the electric and magnetic field decouple and the near field is purely electrical with zero rotation. It was shown in this case that $G_1 = \int_{AllSpace} \varepsilon \vec{E}^2 dV = 0$ [18] and (60) simplifies to

$$\frac{\Delta\tilde{\omega}_{Static}}{\omega} = -2\frac{\Delta n}{n}\frac{\int_{V_s} \varepsilon \vec{E}^2 dV}{\int_{AllSpace} \frac{\partial \omega \varepsilon}{\partial \omega}\vec{E}^2 dV} = -2\frac{\Delta n}{n}\tilde{C}_{static} \tag{61}$$

where C now is the ratio between the energy stored in the sensing volume and the energy stored in the whole near field. In contrast to the energy interpretation of (53) the quasi-static frequency shift of a dispersive resonator is a factor of two stronger. This effect can be understood as a positive feedback from the resonators dispersion. When the frequency decreases, the permittivity of the resonator decreases as well and thus the electric field is pushed out of the resonator which leads to an additional shift.

If the resonance is at a frequency where the metal has low loss it has been shown by Wang and Shen [18] that generally the ratio between field energy in the metal $U_m$ and in the dielectric surrounding $U_d$ is always bigger than one and is given by

$$U_m/U_d = \frac{d(\omega\varepsilon')}{d\omega}\bigg/-\varepsilon' \tag{62}$$

where $\varepsilon'$ is the real part of the permittivity of the metal. With the energy interpretation of (56) C for the case of a bulk change of refractive index can be written as

$$C_{static} = \frac{U_d}{U_m + U_d} = \frac{1}{1 + U_m/U_d} = \frac{\varepsilon'}{\varepsilon' - \frac{d(\omega\varepsilon')}{d\omega}} \tag{63}$$

which is then always smaller than ½. Regardless of any properties of the resonator and the analyte this leads to an upper bound of the frequency change:

$$\Delta\omega_{Max} \leq \omega/n \qquad (64)$$

### 4.3.6. Comparison with Literature

With the frequency shift (61) the perturbation theory can be compared to the analytical theory for bulk sensitivity by Miller and Lazarides [11]. They used a linear approximation for ε'=mλ+ε0 where lambda is the wavelength of light and calculated the wavelength shift of a plasmonic resonance based on the scaling of the quasi-static resonance condition:

$$\frac{d\lambda}{dn} = \frac{2\lambda}{n} + \frac{2\varepsilon}{nm} \qquad (65)$$

Using their approximation for ε' and inserting it into (61) the wavelength shift from perturbation theory turns out to be the same expression. This shows that their theory is contained as a special case in the theory developed here.

Interestingly this can be also shown in a different way, avoiding the scaling arguments of Miller and Lazarides and deriving the quasi-static expression directly from the constitutive equation of quasi-statics. From $\int_{AllSpace} \varepsilon \vec{E}^2 dV = 0$ it follows that [18]

$$\varepsilon_M = -\frac{\int_D \varepsilon_D \vec{E}^2 dV}{\int_M \vec{E}^2 dV} \qquad (66)$$

Where D and M stands for dielectric and metal. Now inserting a perturbation $\varepsilon_D = \varepsilon_0 + \eta \varepsilon_1$ using $\varepsilon_M = \varepsilon_M(\omega(\eta))$ and differentiating (66) leads to

$$\frac{\partial \varepsilon_M}{\partial \omega}\frac{\partial \omega}{\partial \eta} = -\frac{\int_D \varepsilon_1 \vec{E}^2 dV}{\int_M \vec{E}^2 dV} \qquad (67)$$

now inserting $\partial \varepsilon_M / \partial \omega = (\partial(\omega \varepsilon_M)/\partial \omega - \varepsilon_M)/\omega$ one gets

$$\frac{\partial \omega}{\partial \eta} = -\omega \frac{\int_D \varepsilon_1 \vec{E}^2 dV}{\int_M (\partial(\omega \varepsilon_M)/\partial \omega - \varepsilon_M) \vec{E}^2 dV} = -\omega \frac{\int_D \varepsilon_1 \vec{E}^2 dV}{\int_{AllSpace} \partial(\omega \varepsilon)/\partial \omega \vec{E}^2 dV - \int_{AllSpace} \varepsilon \vec{E}^2 dV} = -\omega \frac{\int_D \varepsilon_1 \vec{E}^2 dV}{\int_{AllSpace} \partial(\omega \varepsilon)/\partial \omega \vec{E}^2 dV} \qquad (68)$$

which is the same as (61). Furthermore it shows that the Theory of Miller and Lazarides can be extended for sensing events different than changes in bulk refractive index simply by adding a shape factor which is a function of analyte size, position and shape and is just the ratio $C_{static}$ for any analyte to that for bulk changes.

## 4.3.7. An example of frequency shifts of a quasi-static resonator

To illustrate the above, a simple example is shown. It consists of an gold nanorod which is 40 nm long and 10 nm in diameter with spherical end caps. It is excited with the E-field polarization along its elongated axis (see figure 23) and has a shape parameter L (although it is not really an ellipsoid) of 0.1 implying a quasi-static resonance around $\varepsilon_M$=-12.5 [2], which is for gold fulfilled at a resonance wavelength of 640 nm. As an analyte a change of the bulk refractive index and the attachment of a small sphere to the end of the resonator is chosen. This geometry was selected for several reasons. First because of its radial symmetry it can be calculated as a 2D problem greatly speeding up computation time and accuracy. This problem is calculated with the FEM method (JCMwave) where adaptive mesh refinement and error estimation is available. Secondly the size of the rod is chosen so that the problem is guaranteed to be quasi-static so all the above described approximations are perfectly valid. The problem can therefore be used to test the theory and later to study the effects of deviations from the quasi-static regime.

The model was setup in JCMwave with a coarse triangulation of approximately 2 nm mesh size (see figure 23). Adaptive mesh refinement was used to reach an error in the electric energy density smaller than 0.1%. Then the near field and the far field in forward scattering direction 1m away from the structure was calculated. The extinction cross section was then calculated by the optical theorem [16]:

$$\sigma_{Ext} = Q_{ext} A_{Rod} = \frac{4\pi}{k} \text{Im}\{\frac{r}{e^{kr}} \vec{\kappa E}(r)\} \tag{69}$$

where $Q_{ext}$ is the scattering efficiency, A is the projected area of the rod, r is the distance where the field is measured and κ is the polarization vector of the incident light.

The exact frequency shifts were calculated with the same model and the corresponding changes in refractive index. From the near field of the blank rod the integrals of (60) for the perturbation theory were calculated.

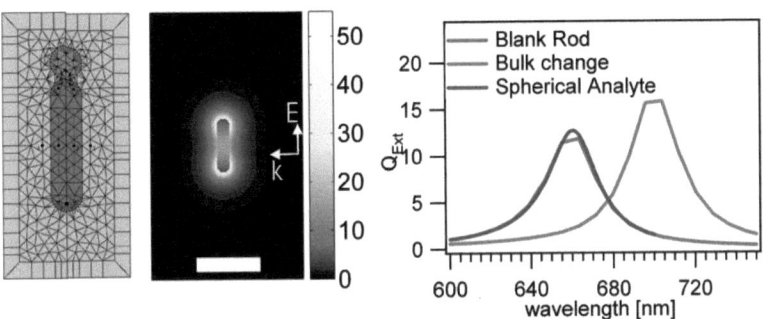

Figure 23: Simulated spectra of a gold nanorod for different analytes. Left: Model and field, the scale bar is 50 nm For symmetry reasons only half of the shown model has to be computed, Right: extinction spectra for the blank rod and a spherical analyte (d = 10 nm $\Delta n = 0.1$) and a bulk change of $\Delta n = 0.1$. The shift of the spectra with the spherical analyte is so small that it is barely visible in the plot.

The results are shown in figures 23 to 25. Figure 24 shows the integrals of (60) as a function of the integrated volume. when compared to the near field in figure 23 it can be seen that convergence is achieved as soon as the integration covers the complete near field (ca. 40 nm distance from the resonator). As expected, the radiation field can be neglected and therefore $G_1$ converges to zero. The same convergence can be shown for the relative wavelength shift $\Delta\lambda/\Delta n$ in figure 25. When converged a shift of 394 nm/RIU is calculated with perturbation theory while the exact calculation yields 411 nm/RIU. This gives an error of the perturbation theory of only 4%

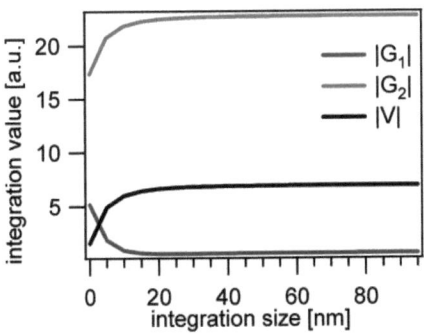

Figure 24: Convergence of the Integrals for the perturbation theory as a function of integration volume (given as the distance from the surface of the resonator).

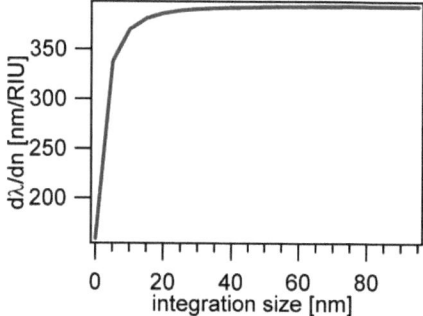

**Figure 25:** Convergence of the wavelengths shift from perturbation theory as a function of integration volume (given as the distance from the surface of the resonator).

The same calculation was done for the spherical analyte placed at the end of the rod with a size d = 10 nm. The resulting shift from perturbation theory is 7.4 nm while the exact calculation yields 7.6nm which corresponds to an error of only 3%. It can be seen that the perturbation theory in the quasi-static regime provides a very exact estimate of wavelengths shifts.

### 4.3.8. Deviations from the quasi-static behaviour.

To study the influence of deviation from the quasi-static regime the same calculations where done for rods of 270 nm and 800nm length and the same aspect ratio. The extinction efficiencies of these rods are shown in figure 26. It can be observed that the spectra shift to longer wavelength due to retardation effects. Also the bigger the rod gets the broader the spectrum gets due to the additional energy loss by radiation. For the biggest rod the extinction efficiency is going down but is still well above the large scatterer limit of 2 which is expected because the rod is still smaller then the exciting wavelength.

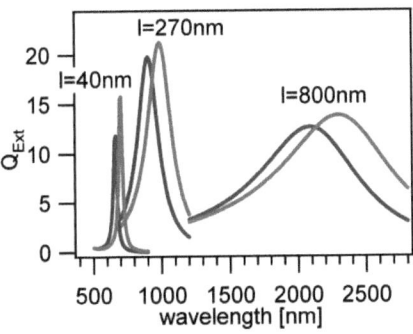

**Figure 26: Extinction efficiencies for rods of different length. Blue is for an ambient refractive index of 1 while red is for an ambient refractive index of 1.1**

For the 270 nm rod the convergence of G1, G2 and V is shown in figure 27. It can be seen that G1 now does not converge to zero anymore but to a finite value. Furthermore there is now an oscillation on the integrands visible, which is due to the radiated wave which cannot be neglected anymore. From the error considerations above it was shown that because the far field does not contribute to the frequency shift, the perturbation result should be sufficiently converged when the integration is done only over the near field. A simple expression providing an estimate of the near field expansion is to define the near field as the region around the resonator where the field is enhanced more than two times compared to the incident wave. Other definitions exist and are more physical (for example taking all field where the time averaged Poynting vector is mostly imaginary [16]) but this one is particular easy to evaluate. For the resonator considered here this area extends 130 nm into the resonators surrounding. Integrating to this leads to a bulk shift of 835nm/RIU (see figure 28). The exact value is calculated to be 803nm/RIU, so the perturbation result has an error of 4% which is still reasonable.

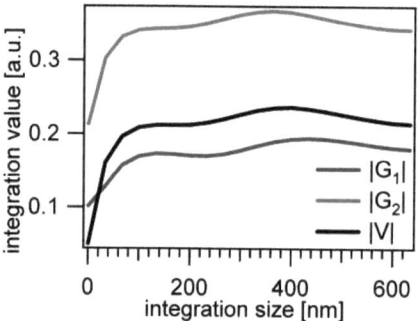

**Figure 27:** Convergence of the Integrals for the perturbation theory as a function of integration volume (from the surface of the resonator) for a rod length 270 nm.

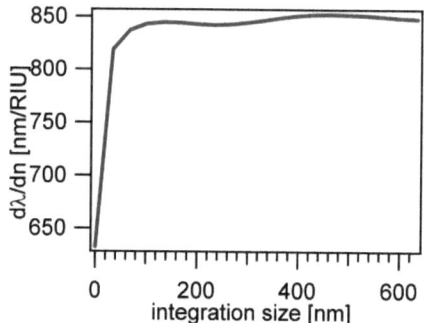

**Figure 28:** Convergence of the wavelengths shift from perturbation theory as a function of integration volume (from the surface of the resonator) for a rod length of 270nm.

Going to even larger rods of 800 nm length, the integrands G1 and G2 can even become comparable in magnitude as can be seen from figure 29. In this case radiation is so strong that it cannot be neglected anymore and leads to a strong oscillation over integration size (figure 25). Integration over the near field as described above and neglecting the surface term leads to a shift from perturbation theory of 1800 nm which, compared to the 2100nm of the exact solution is an error of 10% which still is acceptable, especially when considering that the Q of this resonator is only 3. From this example it can be concluded that perturbation theory is in every practically interesting case suitable to describe the response to changes in refractive index accurately. In cases where the shifts are small and full 3D calculations are required due to a lack of symmetries it will be even superior to exact calculations as these will need a very high accuracy. Even more interesting, with an analytical description of the frequency shifts only from physical properties of the resonator, an analytical expression for the FOM can now be written down.

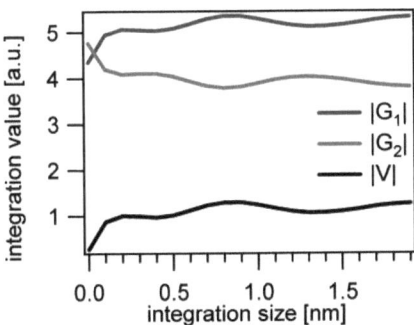

Figure 29: Convergence of the integrals for the perturbation theory as a function of integration volume (from the surface of the resonator for a rod length of 800 nm

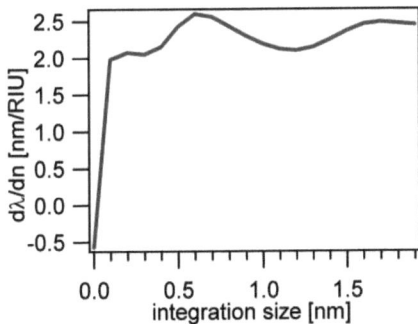

Figure 30: Convergence of the wavelengths shift from perturbation theory as a function of integration volume (from the surface of the resonator) for a rod length of 800 nm.

## 4.4. The analytical expression for the FOM

Now that expressions for the response $\Delta\omega$ and the measurement uncertainty $\Delta\omega_{Un}$ are available, an analytical expression for the FOM can be written down. In the signal noise regime it is given by inserting equations (56), (45) and (47) into (41):

$$FOM \propto \frac{\Delta\omega}{\Delta\omega_{un}} = \frac{\Delta n}{n} \cdot \frac{\sqrt{P \cdot T \cdot I \cdot t}}{f} \cdot \frac{\omega \cdot C\sqrt{\sigma}}{\Gamma} \tag{70}$$

This FOM was divided into three parts which reflect the influence of the analyte itself, the experimental setup and the plasmonic resonator. It can bee seen that these three parts contribute multiplicative to the FOM. From the analyte itself the FOM is influenced via the refractive index contrast, which could be improved by labelling the analyte, for example with metal nanoparticles. From the instrumental side it can be improved by increasing the intensity and transmission trough the microscope or increasing the integration time. The FOM for particle optimization in the signal noise regime becomes

$$FOM^{SN} = \frac{\omega}{\Gamma} C\sqrt{\sigma} \tag{71}$$

the expression $\omega/\Gamma$ equals the quality factor Q of a resonator [16] so this equation can be written as

$$FOM^{SN} = QC\sqrt{\sigma} \tag{72}$$

For the other noise cases the same derivation leads to results which differ only in the weight of the cross section to the FOM. In the background noise regime the result is

$$FOM^{BN} = QC\sigma = FOM^{SN}\sigma^{0.5} \tag{73}$$

The instrument noise case can be considered as an effect which lowers the effective cross section of the particle. Because this lowered cross section does not depend on the resonator it drops out and the FOM becomes

$$FOM^{IN} = QC = FOM^{SN}\sigma^{-0.5} \tag{74}$$

The general expression for the FOM is therefore

$$FOM = QC\sigma^{\alpha} \tag{75}$$

where $\alpha$ is chosen according to the dominating noise case.

Equation (75) is the general result of this chapter. It shows which properties an optimized plasmonic resonator should fulfil. It should have a high quality factor, strong field confinement to the analyte volume and, depending on the noise case a high absorption or extinction cross section. Clearly the confinement factor C depends strongly on the type of analyte which is to be detected. As this FOM is completely general it allows now to compare different sensor concepts which will be done in the next chapter.

## 4.5. The FOM for different sensor concepts

Equation (75) allows now to compare different sensor concepts which are conceptually different in the quantities that constitute the FOM. Several combinations are possible. A photonic mode which is not bound to a surface but bound to a lossless volume can have very high quality factors because of absent loss mechanisms but its modal volume is comparable or bigger than half the wavelength which gives a low confinement C. The extreme example for this concept is a whispering gallery mode resonator which can have Qs in the order of $10^8$ [42, 43] but have an extreme small confinement. Still it is possible to detect single molecules with such a resonator. Plasmonic resonators are on the other side of the spectrum of concepts and employ a fairly low Q but very high confinement. Photonic crystals are somewhere in between and can have high Q and modal volumes in the order of half the wavelength. The contributions of Q,C and $\sigma$ to the FOM for different sensing concepts are summarized in table 1 and sketches of these concepts are shown in figure 31. The decision for a concept cannot only be based on its sensor properties but is often additionally constrained by further requirements like chemical stability, the possibility to use surface chemistry for functionalisation or the ease of use. Plasmonic resonators are often advantageous under such constraints. When made of gold they are very easy to functionalize with sulphur chemistry and are very stable against chemical influences. Another advantage is that because of the high confinement it is possible to measure in dense solutions like blood which would be impossible with sensors like whispering gallery mode resonators due to their huge modal volume. Also spectroscopy on plasmonic resonators is easy to do and does not require special temperature stabilized light sources or fiber coupling to the resonator.

| Sensor type / Quantity | Q | C | $\sigma$ |
|---|---|---|---|
| Plasmonic resonators | low | Very high | Medium |
| Propagating plasmons | low | low | very high |
| Whispering gallery modes resonators | very high | very low | low |
| Photonic crystal | high | medium | medium |

**Table 1: Comparison of the contributions to the FOM for different sensor concepts.**

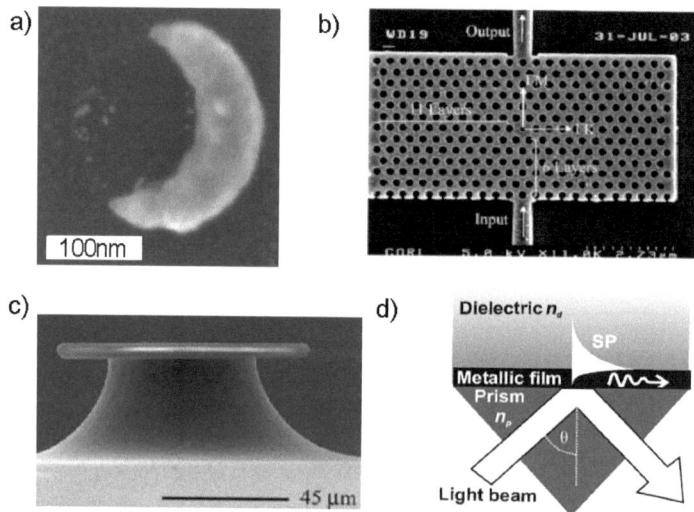

**Figure 31: Transducers in different refractive index sensing concepts. a) LSPR b) Photonic crystal [44] c) whispering gallery mode resonator [42] d) propagating surface plasmon sensor [45]**

It would now be interesting to see if it is possible to combine advantageous properties of high Q and high C in one resonators. From table 1 it seems that no such resonators exist and high C has to be achieved with corresponding losses in Q. For plasmonic resonators the reason is clear: as the mode is bound to the surface of a lossy metal particle when going to smaller sizes and high C the absorption losses in the metal remain and in the quasi-static case are independent of the resonator shape and size. The only way to overcome these intrinsic losses would be to compensate them by using an additional layer of a gain material, like organic dyes or quantum dots. It was shown [18] that Q in the presence of a gain material, covering the whole modal volume is given by

$$Q = \left(\frac{\varepsilon_m^{''}}{|\varepsilon_m^{'}|} - \frac{\varepsilon_g^{''}}{|\varepsilon_g^{'}|}\right)^{-1} \frac{\omega\left(\frac{d\varepsilon_m^{'}}{d\omega} + \frac{|\varepsilon_m^{'}|}{\varepsilon_g^{'}}\frac{d\varepsilon_g^{'}}{d\omega}\right)}{2|\varepsilon_m^{'}|} \quad (76)$$

Where $\varepsilon_g$ is the permittivity of the gain material. This Q diverges when

$$\frac{\varepsilon_m^{''}}{|\varepsilon_m^{'}|} = \frac{\varepsilon_g^{''}}{|\varepsilon_g^{'}|} \quad (77)$$

This means that such a particle becomes an oscillator equal to a laser. The concept of a lasing surface plasmon mode has recently gained much interest based on the fundamental theoretical work of Bergman and Stockman [46] who called this concept a SPASER (acronym for surface plasmon amplified emission of radiation). Recently it was demonstrated experimentally [47]. This shows that it is in principle possible to achieve any desired Q with this advanced concept. Practically it will be difficult to built such a structure due to the fact that all space covered by the gain medium is lost for sensing thus C will maybe lowered again.

## 4.6. The FOM for different types of analytes

The following section therefore discusses three experimentally important cases of analytes, namely the detection of changes in bulk refractive index, the adsorption of thin dielectric layers to the surface of the resonator, and the change of refractive index in a very small spherical volume close to the resonator. A standard way of comparing different resonators in literature is to compare their response to changes in bulk refractive index divided by their line width. For this case, a $FOM_{B,Lit} = m/\Gamma$ is introduced in literature [8] with the bulk sensitivity factor $m = d\omega/dn$, which, from equation (60), equals

$$FOM_{B,Lit} = \frac{m}{\Gamma} = Q\frac{C}{n} \qquad (78)$$

which, by comparison with equation (75) with α=0 can be identified as the expression for changes in bulk refractive index in the instrument noise regime up to a constant prefactor $n^{-1}$.

The second frequently encountered situation is the detection of a thin layer which adsorbs on the resonator surface. Here, one has to specify a fixed layer thickness d for which the FOM is evaluated. For very thin layers the field does not vary strongly along a path perpendicular to the resonator surface inside the layer and FOM becomes a linear function of d. In this regime one may alternatively state a $FOM_L$ that is normalized by this value

$$FOM_L = \frac{\omega}{\Gamma}\frac{C}{d}\sigma^\alpha = \frac{\omega}{\Gamma}C_L\sigma^\alpha \qquad (79)$$

avoiding the choice of an arbitrary d by definition of $C_L=C/d$. This equation shows the importance of a field which is strongly confined to the surface. For this situation, the connection between the FOM for layer and bulk sensing should be discussed. It is based on the empirical expression for layer sensing [8, 33]

$$\Delta\omega = m\Delta n(1 - e^{-2d/l_d}) \qquad (80)$$

with an effective decay length $l_d$ which yields for very thin layers

$$\frac{\Delta\omega}{\Delta n} = \frac{2m}{l_d}d \qquad (81)$$

A direct comparison to the established definition of the FOM [8] is possible if we introduce a shift factor for layers, $m_L = 2m/l_d$. in analogy to m used for bulk. Then, we arrive for the instrument noise regime ($\alpha=0$) at a FOM for layers which is formally equal to the established definition for bulk refractive index changes, $FOM_B$ (equation (78)) with m being replaced by $m_L$. This expression is identified with the FOM for layers proposed here (eqn. 79) up to a factor $n^{-1}$, if the instrument noise ($\alpha=0$) is considered

$$FOM_{L,Lit} = \frac{m_L}{\Gamma} = \frac{\omega}{\Gamma}\frac{C}{d}\frac{1}{n} = \frac{1}{n}FOM_L^{IN} \qquad (82)$$

Hence for layer sensing a modified shift factor should be used that takes into account the typical decay length of the near field: Highly confined fields are advantageous. This intuitively clear modification implies that resonators which have been optimized for bulk refractive index sensing will in general not be the ideal choice for the sensing of thin layers and vice versa.

In order to define a FOM for single-object sensing, both shape and position of the analyte under study must be specified. The latter quantity can be eliminated if one assumes that the analyte is placed such that the FOM is maximized. An easy geometry to consider is a small sphere with a radius r. Very similar to the situation for layers, for sufficiently small r FOM(r) will become linear in $r^3$ as soon as the field is approximately constant over the sensing volume and we may define

$$FOM_S = \frac{\omega}{\Gamma}\frac{C}{r^3}\sigma^a = \frac{\omega}{\Gamma}C_S\sigma^a \qquad (83)$$

In this regime, the exact shape of the sensing volume does not matter anymore.

## 4.7. Scaling of the FOM

In this section the scaling of the FOM with resonator size and shape will be discussed. This is interesting for two reasons. First, numerical calculations, especially for resonators which lack symmetries and have to be calculated in 3D, are usually computationally demanding and take long times. For this reason it is desirable to know a priory which size and shape an optimal resonator should roughly have and take this knowledge to restrict the parameter space for optimization. Second the scaling shows which of the different optimization goals has which weight, and therefore allow also conclusions how to design the experimental apparatus.

Like always completely general statements are not possible to give, because of the richness of possible solutions to Maxwell's equations. Quasi-statics will allow very general statements because in the quasi-static regime the resonator properties are mainly dependent on the material properties. Small deviations from quasi-statics can be included heuristically but interpretations have to be done with care.

In the quasi-static regime Q is only dependent on the metal loss. In this regime like for C an analytical expression for Q can be found [18]

$$Q_{static} = \frac{\omega d\varepsilon'/d\omega}{2\varepsilon''} \qquad (84)$$

where $\varepsilon''$ is he imaginary part of the metal permittivity. C for every type of analyte can be written as $C_A = C_{static} f(r, V_a)$, where f is the weighted fraction of the maximum possible sensing volume, which is occupied by the analyte (which means f<=1). It is dependent of the characteristic size of the resonator, the type of analyte and the volume occupied by the analyte. Because of the inhomogeneity of the near field it also depends on the position of the analyte, if the analyte does not cover the whole sensing volume. The scaling f with the size of the resonator is dependent on the analyte. For a bulk refractive index change it is constant while for a layer it scales with the film thickness to particle size which is proportional $r^{-1}$ and for a small spherical sensing volume is scales with $r^{-3}$.

Furthermore for the cross section in the quasi-static limit only dipole absorption and radiation is relevant. It can be written as

$$\sigma = \frac{P}{I} \qquad (85)$$

where P is the absorbed or radiated power and I is the incident intensity. $\sigma$ for a dipole is given as

$$\sigma_{Scat} = \frac{k^4}{6\pi\varepsilon_0^2}|\alpha|^2 \qquad (86)$$

if scattering is measured or

$$\sigma_{Abs} = \frac{k}{\varepsilon_0}\operatorname{Im}\{\alpha\} \qquad (87)$$

for absorption. The quasi-static polarizability $\alpha$ of an ellipsoid along one of its principal axes is given as [2]

$$\alpha = V\varepsilon_0 n^2 \frac{\varepsilon - n^2}{n^2 + L(\varepsilon - n^2)} \qquad (88)$$

where L is a geometry parameter which ranges from zero (an infinite cylinder with incident polarization along it's axis) to one (a flat disk with incident polarization normal to it's end face). It is 1/3 for a sphere. V is the volume of the resonator. To discuss the scaling properties of the FOM Q, C and $\sigma$ have to be inserted into (75). Combining C and Q simplifies to the simple expression

$$Q_{static} C_a = -\frac{\varepsilon'}{\varepsilon''} f(r, V_a) \qquad (89)$$

With 0<f<1 Now instead of writing down the complete expression for (75) it is useful to see how it actually looks like. Q C is given by a dimensionless quantity f which depends on the analyte and size of the resonator and material properties of the resonator. σ depends on the size, material properties of the resonator and the geometry factor L of the resonator. In the quasi-static case the resonances occur where the denominator of the polarizability is minimal. This gives a resonance condition [2]

$$n^2 + L(\varepsilon' - n^2) = \frac{1-L}{L}n^2 + \varepsilon' = 0 \qquad (90)$$

from which follows that for a given surrounding of the resonator and resonator material the resonance wavelength and material properties are given by L. It follows that the FOM in the quasi-static limit for a given analyte will generally scale as

$$FOM \propto r^\beta f(L) \qquad (91)$$

where β is dependent on the dominant source of noise and the type of analyte. This result shows that size and shape can be optimized separately in the quasi-static regime.

|  | bulk $FOM \propto C_{bulk} \propto r_{mp}^0$ | layers $FOM \propto C_{layer} \propto r_{mp}^{-1}$ | particles $FOM \propto C_S \propto r_{mp}^{-3}$ |
|---|---|---|---|
| Instrument Noise $FOM \propto \sigma^0 \propto r_{mp}^0 \left(r_{mp}^0\right)$ | 0 | -1 | -3 |
| Signal noise $FOM \propto \sigma^{0.5} \propto r_{mp}^{1.5} \left(r_{mp}^3\right)$ | 1.5(3) | 0.5(2) | -1.5(0) |
| Background noise $FOM \propto \sigma^1 \propto r_{mp}^3 \left(r_{mp}^6\right)$ | 3(6) | 2(5) | 0(3) |

**Table 2: Dependence of the scaling exponent β on resonator size for different dominating sources of the noise and different analyte geometries. In each entry, absorption and extinction cross sections are considered, the latter are given in brackets.**

The dependence of β for different types of analyte and noise is given in table 2.

For bulk refractive index measurements, C is constant and therefore in the signal noise regime large resonators should be preferred for their large scattering and absorption cross sections. In the instrument noise regime the FOM is independent of particle size

For very thin layers, C scales with $r_{mp}^{-1}$ and thus increases with decreasing resonator size. In the instrument noise regime, small particles are advantageous since their confinement is increased and the cross section has no effect on the FOM. Though, including the scaling of the cross sections in the signal noise regime, $FOM^{SN}$ scales with $r_{mp}^{0.5}$ and $r_{mp}^{2}$ for scattering and extinction experiments, suggesting that increasing resonator size is advantageous. The background noise regime favours large resonators even stronger. Remarkably, depending on the dominating source of noise, either increasing or decreasing particle size will improve the performance of the sensor. As a consequence, the true optimum will be found right at the resonator size where the transition between the instrument noise regime and the signal or background noise occurs. This can be understood as a change in the weight of the two different optimization goals, first maximize the frequency shift and second minimize the detection uncertainty. As long as the instrument noise dominates the primary goal is to increase the frequency shift which is determined by the confinement, while in the signal noise regime minimizing the detection uncertainty is more important. If we finally consider attachment of a small dielectric sphere, the confinement scales with $r_{mp}^{-3}$. As a consequence small resonators are advantageous as well if the instrument noise dominates. For scattering the $FOM^{SN}$ is independent on resonator size but in extinction experiments it scales with $r_{mp}^{-1.5}$ and small resonators may be advantageous in the signal noise regime. Again, background noise strongly favours large resonators. Similar to layer sensors, the optimum resonator will correspond to the transition between two different noise regimes.

It has to be pointed out that this analysis is only valid for small radiation losses, hence for structures so small that absorption losses dominate over radiation losses. For bigger structures Q will be dominated by radiation losses which are proportional to $r^6$. In this regime small resonators are better in all cases. An exception can be higher order resonances. There exist resonances which lack a strong dipole moment, and therefore do not radiate strongly. The Q of these resonances can usually not be found by simple models and numerical calculations are required.

The second quantity to scale is the function f(L). In the case of a layer or spherical detection volume it is also dependent on the analyte position which makes it difficult to evaluate. Usually numerical calculations have to be used. It is known that edges and corners at the resonator surface lead to a high field enhancement which enhances the confinement and therefore the sensitivity. Because of this resonators with small features will be advantageous. In the quasi-static case and changes of bulk refractive index f(L) can be calculated only from the dielectric properties of the resonator and the resonance condition (90). For this calculation, (75) is used and the quasi-static expression (89) for CQ and inserted. For the case of a scattering measurement and intensity noise as the dominating noise this gives the expression

$$f(L) = \frac{k^2}{2\sqrt{6\pi}} Abs[\frac{n^2(n^2-\varepsilon)}{L\varepsilon-(L-1)n^2}]\frac{\varepsilon'(L)}{\varepsilon''(L)} \qquad (92)$$

ε is a function of L because of the resonance condition (90). For the case of a gold resonator this function is shown in figure 32.

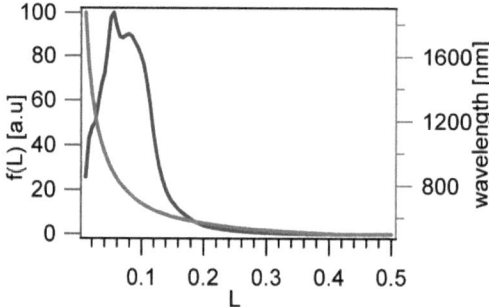

**Figure 32: f(L) for a gold resonator. left axis: f(L), right axis: resonance wavelength as function of L**

It can be seen that there is a maximum around L=0.1 which corresponds to an elongated ellipsoid excited with a polarization along its longer axis. The reason for this maximum is that for smaller L's the wavelength of the resonance wavelength will increase shifting to regions where the absorption in the metal is lower. This increases C and Q as well so the FOM will increase. When the wavelengths shifts further the scattered intensity will go down as it is proportional to $\lambda^{-4}$. When this effect dominates the FOM will go down again. Practically the resonance wavelength should stay below 1000 nm because of the need of well corrected microscope objectives and affordable, sensitive detectors. At L=0.1 the resonance wavelength is around 800 nm, which makes single particle measurements practical. When deviations from quasi-statics occur usually resonances shift fast into the infrared. Very coarsely optimal resonators will therefore be nearly quasi-static and have resonance wavelengths from 750 nm to 1000 nm.

# 5. Characterization of the crescent shaped particles

## 5.1. Introduction

Now that the tools are available, crescent shaped particles which were a main part of this work will be analyzed theoretically. Crescent shaped particles can be produced in several ways: With nanosphere template lithography, described earlier, or with direct electron beam writing. The first method has the advantage to yield much smaller features as electron beam writing is typically limited to feature sizes bigger than 20 nm. The fabrication of crescent shaped particles was described by Rochholz [14] who also investigated experimentally the extinction properties as a function of geometry. Rochholz also tried to describe the effects observed by a simple model and used numerical simulations of very simplified 2D structures. It was not possible in this model to understand the resonances quantitatively and predict sensing properties. Much ambiguity remained on the assignment of resonant peaks to different order resonances. The goal of this chapter is therefore twofold: First the resonances of the crescent shaped particles have to be modelled and their resonances understood in a way that makes it possible to map theoretically obtained and experimental results unambiguously. Second based on the understanding of the bare crescents resonances the sensing properties of the crescent resonators are investigated.

The chapter is structured as follows: First the geometrical description of a crescent shaped resonator is shown. Then simplified description of the crescent as a simple metal rod which is bend is introduced. It is investigated which effects can be described with this model and which effects require numerical simulations. It is shown that even in the simplified model basic sensing properties can be explained in a perturbative approach (mode coupling theory). Then the numerical simulation of such a bend rod is shown and the limits of the simple rod model are studied. Next the sensing properties are investigated. There are two relevant sensing cases which will be considered: the growth of a dielectric layer on the crescent (which contains as a special case the bulk refractive index change) and the change of refractive index in a spherical volume close to it. Where experimental data was available prior to this work it will be repeated here and compared to the theoretical results. Then the sensing properties as a function of crescent size and fabrication parameters are investigated.

## 5.2. Parameterization of the crescent model

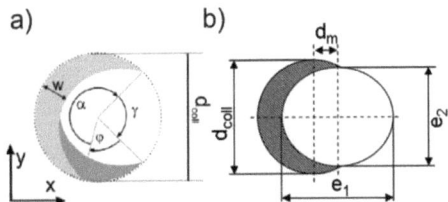

**Figure 33:** a) Sketch of the geometry of the crescent fabricable by nanosphere lithography on a glass support b) Construction of the crescent in the simulation

Figure 33 shows a sketch of such a resonator as it can be experimentally realized by colloidal lithography [13].

For an ideal process its 3-dimensional shape is described in a unique way by four parameters: The evaporation angle Θ determines the width w while the diameter $d_{coll}$ is determined by the size of the colloids used as a mask. Furthermore, the thickness t of the structure in z-direction is determined by the amount of evaporated metal and, for double evaporation, additionally the turning angle φ can be adjusted to tune the opening angle γ.

For the simulations the particle shape was approximated by a circle which is cut by an ellipse (figure 33b). The parameters to describe this geometry are obtained by a fit to scanning electron micrographs of fabricated crescents. This representation of the geometry avoids hard edges along the inner perimeter of the crescent and is described by a simple equation which facilitates the setup of the simulation. Vertical sidewalls are assumed in the present model although fabrication normally leads to tilted sidewalls. This simple shape reproduces well the behaviour of real structures, while calculation of more realistic shapes show only minute differences [34].

The observed resonances are named according to their polarization as "C" or "U" resonances. A "C" resonance corresponds to an excitation E-field polarization parallel to the y-axis of figure 33 while a "U" resonance is excited with polarization parallel to the x-axis. The resonances are named in ascending order starting at $C_1$ or $U_1$ for the resonance with the longest wavelength [13].

## 5.3. Simple Model

A very simple model to understand the basic behaviour of the crescent shaped particle is to consider it as a rod with rotational symmetry along one axis and a homogeneous diameter. To get the crescent shape the rod is then bend with some bending radius. It is known that a metal rod supports propagating waves along its axis z with a propagation constant $n_{eff}$ according to:

$$E(r,\varphi,z) = E(r,\varphi)e^{ikn_{eff}z} = E(r,\varphi)e^{i\beta z} \tag{93}$$

where k is the free space wave vector. The propagation constant $n_{eff}$ can be calculated from waveguide theory, see for example Novotny [48] who used this theory to derive scaling laws for optical antennas. This propagating wave is then reflected at the tip and forms a standing wave at the rod. The condition for resonance is that the phase accumulated in one round trip equals a multiple of 2π.

$$\frac{2\pi}{\lambda_R}n_{eff}2l + 2\varphi_{Tip} = m2\pi \tag{94}$$

Here $\varphi_{Tip}$ is an additional phase shift at the end of the rod, $\lambda_R$ is the resonance wavelength and m is an integer number.

In these standing waves, charge accumulates at different positions depending on the order of the resonance. The resonance can be excited via its dipole moment which is given by

$$\vec{p} = q\sum_{i}^{m+1}(\pm)\vec{r}_i \qquad (95)$$

where q is the accumulated charge, r is the position of the charge and the sign is given by the sign of the charge at a fixed instance in time (the charges oscillate as exp(-i$\omega$t)).

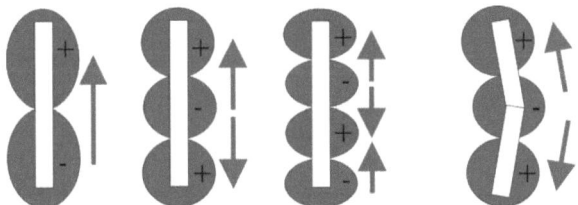

**Figure 34: Sketch of the charge distribution of the first three resonances of a metal rod. The arrows indicate the dipole moments which contribute to the total dipole moment of the rod. The right image shows the second order resonance of a slightly bend rod which results in a nonzero dipole moment.**

This simple model already gives a qualitative explanation of all observed features of the crescents. First depending on the length of the rod multiple resonances are possible. figure 34 shows the charge distribution of the first three resonances along with the corresponding dipole moments. It can be seen that for the second order resonance the dipole moments cancel out and hence the net dipole moment is zero. The same is true for higher order resonances with an even m. Consequently these modes cannot be excited simply with an incident plane wave and are so called dark resonances. When the rod is bent, however a dipole moment perpendicular to the rod remains and the even modes can be excited.

Also the detection of changes in refractive index can be described in this model. When the effect of the tips is neglected, the change of refractive index leads to a change in the propagation constant $n_{\text{eff}}$ and therefore a change in the resonances wavelength. The slope of this change is given by:

$$\frac{d\lambda}{dn_{\text{eff}}} = \frac{2l}{m} \qquad (96)$$

The longer the rod and the smaller the order of the resonance the bigger this slope will be. One can also insert in this expression the resonance condition equation (94). When the influence of the tips can be neglected this leads to the particular simple expression

$$\frac{d\lambda}{dn_{eff}} = \frac{\lambda_R}{n_{eff}} \qquad (97)$$

This shows that the slope is only dependent on the resonance wavelength. When the bending gets stronger, additional effects due to the coupling of the electrical fields from the ends of the rod occur. This will be studied in the next chapter. Also the problem of calculating the sensitivity is now reduced to finding the change of $n_{eff}$ which is only a 2D problem and can be analytically solved by mode coupling theory. From this very simple model already sensitivities and penetration depths from the various modes can be calculated.

A change of the refractive index $\Delta n$ within the volume where the electric field E is not vanishing will lead to a change $\Delta n_{eff}$. This change is in first approximation obtained by mode coupling theory [49]

$$\Delta n_{eff} \propto \int |E|^2 \Delta n^2 dV \qquad (98)$$

Here the integral is to be taken over the entire space. If, as it is the case here, a material with a fixed refractive index is added so that it occupies some volume in the nearby area, for a resonator with a given length $l$ a change in $n_{eff}$ will directly lead to a change in resonance wavelength, then

$$\Delta\lambda \propto \Delta n^2 \int_{V_{oc}} |E|^2 dV \qquad (99)$$

Here the integral is to be taken over the volume ($V_{oc}$) that is occupied with the high refractive index material. This theory is only strictly valid for real dielectrics or perfect metals, which is not the case for gold in the visible range. Furthermore it is assumed that the perturbing change in the refractive index only leads to infinitesimal changes of the field, which is only true for small dielectric contrasts $\Delta n$.

For the cylindrical rod the field distribution can be calculated analytically. In this case, the radial electrical $E_r$ field outside the rod is given by

$$E_r \propto H_1^{(1)}(k_r r) \qquad (100)$$

where $H_1^{(1)}$ is a Hankel function of the first kind, and r the distance measured from the center of the rod. The $E_r$ component of the electrical field is the normal component and dominant in the near field. This leads to

$$\Delta\lambda(d) \propto \int_R^{R+d} r \left| [H_1^{(1)}(k_r r)]^2 \right| dr \qquad (101)$$

with $R$ being the radius of the rod and $d$ the thickness of the layer. The crescent is modelled as a rod with a R = 20 nm corresponding to half the height of the crescents. The resulting normalized $\Delta\lambda_i(d)$

curves are displayed in figure 35. A typical decay length ($d_{l,rod}$) was obtained in this approximation by solving the equation $\Delta\lambda(d_{l,rod}) = (1-e^{-1})\Delta\lambda_{max}$, $\Delta\lambda_{max}$ being the shift for an infinitely thick coating.

Experiments where carried out by Bocchio [33] where layers of polyelectrolytes where grown on crescent shaped resonators. From the resonance wavelengths of the uncoated resonator the propagation constants can be calculated with equation (94) when the influence of the crescent tips is neglected. Then decay constants can be calculated and compared to experimental values. The results are shown in table 3.

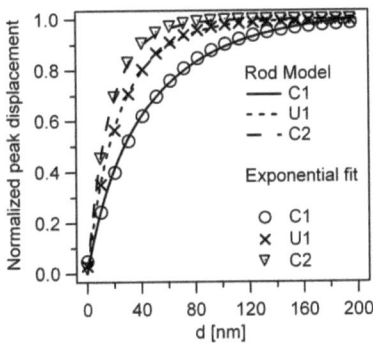

Figure 35: Normalized peak shift, as calculated from the rod model, as a function of the distance to the rod's surface (d).

| Resonance | $C_1$ | $U_1$ | $C_2$ |
|---|---|---|---|
| Wavelength [nm] | 1364 | 897 | 742 |
| $k_x$ [ nm$^{-1}$] | 0.00855 | 0.0171 | 0.02565 |
| $k_z$ [ nm$^{-1}$] | 0.0072 | 0.0156 | 0.0242 |
| $d_{l,exp}$[nm] | 26.2 | 17.2 | 8.5 |
| $d_{l,rod}$ [nm] | 44.5 | 25.5 | 17.6 |

Table 3: Decay lengths for the different modes according to the measurements ($d_{l,exp}$) and the different models ($d_{l,PI}$, $d_{l,rod}$)

The very simple geometrical models reproduce qualitatively the experimentally observed trend of a shrinking modal volume with increasing mode number. In particular, the direct connection to a simple analytical field distribution allows for an intuitive understanding of the observed effects. Several mechanisms may account for the quantitative difference between the models and the experiments. Firstly, the geometrical simplifications made when describing the crescents as rods are significant, in particular, at the tips of the crescents, where a high localization of the field occurs which is not reflected in the models and can lead to a reduced $d_l$. Secondly, the mode coupling theory itself and the description of the electrical near field is only an approximation. Finally, experimental imperfections of the structures and the coating may also play a role, particularly for the first layers.

## 5.4. Simulation

In the previous chapter it was shown that the basic properties of the crescent shaped particles can be understood from a simple model without the need for numerical simulations. Still there occurred deviations which could be attributed to the simplifications in the model. For example the role of the tips can be not understood in a simple rod model. Furthermore from the sensing theory in chapter 4 it has become clear that for an evaluation of the sensing capabilities especially for single molecules or very few molecules the exact field distribution around the resonator has to be known. Because of the lack of symmetries the response of the crescent shaped particle can only be calculated numerically and full 3D simulations are needed. The following chapters will therefore show numerical results, first for the simple bend rod model and then for realistically shaped crescent shaped particles.

## 5.5. Numerical simulation of the bend Rod model

To study the influence of the rod bending on the optical response the model from figure 36 is used. The rod has a rectangular cross section and is positioned on a glass support and described by four parameters, namely the thickness t, length l, the width w, and the bending radius r. To get rid of additional effects caused by sharp tips the end are rounded with a radius of w/2.

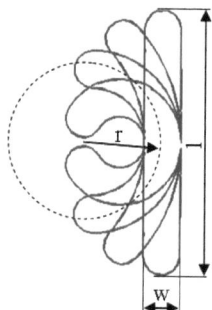

**Figure 36: Parametrisation of the rod**

The response of the rod to a plane wave incident perpendicular to the support and polarized along l and w is calculated numerically with the FEM method. After calculation of the local field the scattered signal in the far field is calculated. The resulting wavelength spectra for a rod with l = 300nm, w = 50nm and t = 40nm are shown in figure 37 for bend radii of infinity (a straight rod) to 58 nm (an almost closed rod). In figure 38 the resonance wavelength of the first two resonances is obtained by fitting the spectra to Lorentzian functions and shown as a function of bend radius.

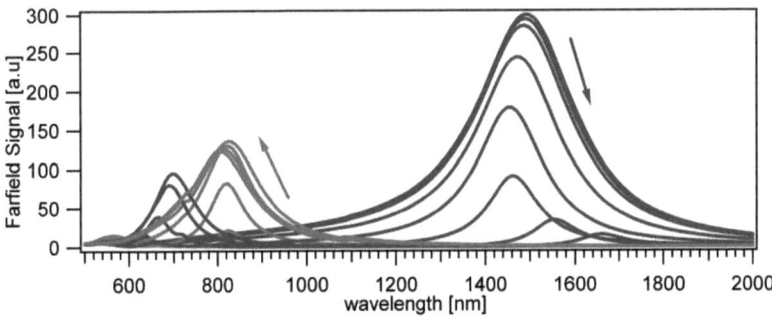

**Figure 37: Spectra of the bend rod for various bend radii. The arrow show the evaluation of the mode when the bend radius decreases**

It can be seen that according to the simple rod model for a straight rod only two resonances with polarization along the rod can be detected. These are the first and third order resonances. When the rod is only slightly bent between these resonances a third one occurs. For small bend radii up to roughly 100 nm, which corresponds almost to a rod bend to a half circle, the position of the resonances doesn't change much. This is the range where the simple model above is valid. The scattered intensity of the second order resonance increases due to the dipole moment increasing with smaller bend radius while the scattered intensity of the second and third order resonance decreases, due to the effective length of the rod for this polarization decreases and with it the dipole moment.

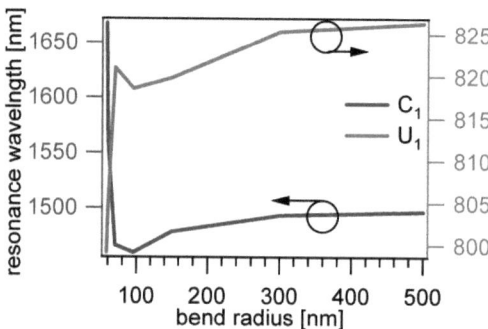

**Figure 38: Resonance wavelength of the bend rod as a function of bending radius for the first two resonances.**

When the bending radius decreases further a small blueshift of the resonances can be observed. This can be understood as a weak coupling between the fields at the tips of the rod and can be qualitatively described by perturbation theory arguments similar to the ones used in chapter 4.3. The shift of the resonance is proportional to the refractive index contrast between air and metal and the fraction of field energy which is affected by the other tip. Because the real part of the refractive index of metal is smaller than that of air, a blueshift occurs.

When the bend radius is further decreased the first order resonance decreases strongly in strength and shifts fast towards the red while the wavelength of the second order resonance shifts only slightly to the blue. For the first order resonance this can be understood in terms of an LC circuit, a concept which was also applied for metamaterials [50]. At the first order resonance the tips of the rod are differently charged and hence behave like the two plates of a capacitor, while the rod itself behaves like an inductance. The resonance frequency of an LC circuit is given by

$$\omega = 1/\sqrt{LC} \tag{102}$$

where C is the capacitance and L the inductance. The capacitance of a plate capacitor is given by:

$$C = \varepsilon\varepsilon_0 A/d \tag{103}$$

where A is the area of the plates and d is the distance of the plates. When the distance of the plates which are in this case the tips decreases the capacitance increases and therefore also the frequency decreases with the square root of the tip distance.

For the second order resonances the tips are charged equally and in a first approximation there is no interaction except for a stronger confinement of the field when the space between the tips decreases. Consequently no strong wavelength shift is observed.

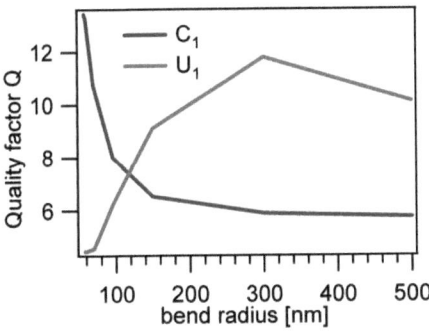

**Figure 39: Quality factor Q of the bend rod as a function of bending radius for the first two resonances.**

Another important quantity to consider is the Quality factor Q. The behaviour as a function of bend radius is shown in figure 39. The Q factor reflects the losses from to mechanisms: radiation losses and absorption losses in the metal of the resonator. For the quasi-static case radiation is negligible and only the absorption losses matter. In this case there exist analytical formulas for Q, which show that Q is only dependent on the resonance wavelength and material of the resonator [18]. When the size of a resonator increases, so that a quasi-static description is not valid anymore, at first it is expected that the Q factor decreases due to additional radiation losses. It was however speculated and later shown by Feigenbaum et al. [51] that the Q factor can also be higher than in the quasi-static case, when the field energy is stored not inside the resonator but in it's near field.

From figure 39 it can be seen that for the first order resonance the Q factor increases strongly when the bend radius gets smaller. For resonance wavelengths below 1200nm the quasi-static Q factor is always smaller than 10. This value is exceeded for bend radii smaller then 100nm, which is when strong interaction between the tips occurs. This shows that in this case the field energy is strongly concentrated in the gap between the two tips, making this case interesting for sensing. Unfortunately the wavelength shifts strongly into the IR, so this resonance is difficult to use for spectroscopy. Also the gaps needed to obtain this effect are very small and not easy obtained practically. For realistic crescents with resonances not too far in the IR this effect is not observed as will be shown later.

For the second order resonance only a decrease of Q is seen when the bend radius decreases. That shows that the increase of radiation is the dominating effect in this case. The Q value is hence always smaller than in the quasi-static case which is 18 in gold at a wavelength of 800 nm. It will be interesting to see if the quasi-static values can be exceeded when the structure is made smaller. This is investigated in chapter 5.7 after the investigation of the sensing properties of a simple crescent shaped resonator.

## 5.6. Sensing properties of a simple crescent shaped resonator

While in the previous chapters it was shown that the crescent shaped particles can be modelled as a bend rod, in this chapter a special case is considered in detail and with realistic shape. This is done for several reasons: Firstly the structure presented here is one of the simplest to produce because it only needs one evaporation step. Second for this particular resonator experimental data is available and also experiments were done in this work (see chapter 6), which show the potential of this structure for single molecule sensing. The results of this chapter were published earlier [32]. Additionally the exact model considered here also accounts for realistic shapes of the tips which is considered to become important for very small sensing volumes.

### 5.6.1. Bare resonator

The particles considered here have the fabrication parameters $\varphi = 0°$, $\Theta = 30°$, $t = 40$ nm, $d_{coll} = 205$ nm. These parameters are used throughout the following discussion. From the calculated extinction cross section, a VIS-NIR absorption measurement can be modelled and compared to measured spectra up to a constant prefactor which is given by the surface coverage. The latter was extracted from SEM micrographs and found to be around 1%. Throughout this chapter all numerical results are obtained with the FDTD method. The comparison of experimental results and a FDTD calculation is shown in figure 40 and is quite good. Below 800 nm where the $C_2$ resonance occurs, deviations are seen which are due to several problems:

Firstly, the dielectric function of gold is not accurately described by a Drude-Lorentz model at smaller wavelength, so the dielectric function of the gold in the simulation deviates from the experiment. Secondly, the regular discretisation which leads to a staircasing approximation of the curved interfaces of the particle affects the smaller wavelengths stronger at a given spatial discretisation. Thirdly, a non-perfect absorbing boundary leads to small computational errors around 800 nm due to resonances in the fictious box used for the simulation. In particular the weak $C_2$ resonance is sensitive to these errors. Beside this there is also inhomogeneous broadening due to the size distribution of resonators in the experiment making the experimentally observed resonances slightly broader then the simulated ones. Also at 1100 nm a small peak can be seen in the experimental spectra but not in the simulated ones. This peak comes from aggregates which form during the fabrication process and was discussed earlier [13].

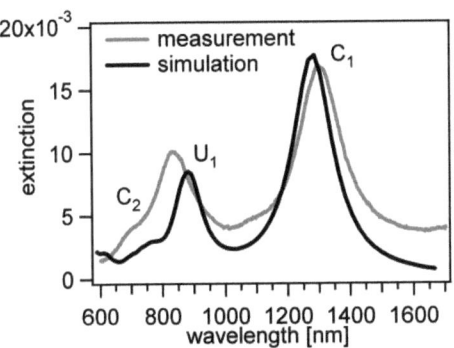

**Figure 40:** Extinction of a layer of isolated crescents

### 5.6.2. Response to attached layers

Attached layers are one of the most important sensing cases. Almost all currently available optical sensors which are based on refractive index sensing use functionalized surfaces where the analyte attaches and forms some type of effective layer. Because of the deep subwavelength structure of this process it can be described by effective material properties even if the layer is not dense and closed. For this case, experimental data is available [33]. In this simulation the response of the crescent to an applied thin dielectric coating with thickness $t_{coat}$ and refractive index $n = 1.5$ was studied. The coating is defined by all space that has a smaller distance to the crescent than the coating thickness and is not part of the supporting substrate (See Inset figure 41b). A change in the bulk refractive index is contained as the special case of an infinitely thick layer. The results are shown in figure 41. The attachment of a thin non-absorbing layer results in a clear red shift of the observed peaks in the spectrum. Lower order peaks shift more than high order peaks which leads to better separation of the $C_2$ from the $U_1$ peak. A general increase of the extinction at lower wavelengths is also observed which can simply be interpreted as an additional Raleigh scattering from the coating. Figure 41b) summarizes the shifts as a function of $t_{coat}$. For a layer thickness below 20 nm, the peak shift is approximately linear in $t_{coat}$, for thicker dielectric coatings the shift saturates.

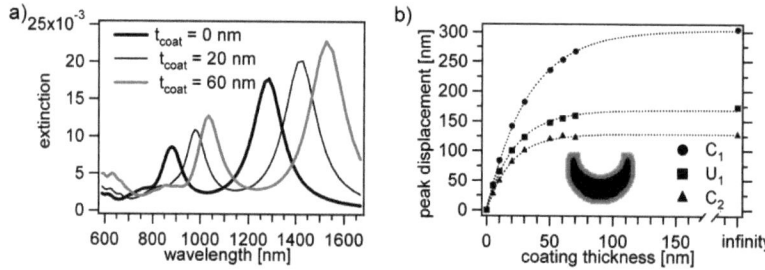

Figure 41: a) calculated extinction spectra for 3 representative $t_{coat}$. b) shift of the resonance peaks when covered by a thin layer of a dielectric with n = 1.5 with varying the coating thickness. The lines are fitted exponential functions. The inset shows a cut through a coated crescent (gray = coating with n = 1.5).

An effective penetration depth of the electric field may be obtained from fitting of the peak displacement curves to an exponential function. The result is shown in table 4. Except for the $C_2$ peak we find very good agreement between the theory and experiment [33].

|  | $C_1$ | $U_1$ | $C_2$ |
|---|---|---|---|
| Simulation | 33 nm | 23 nm | 19 nm |
| Experiment | 29.1 nm | 21 nm | 10 nm |

Table 4: Comparison of simulated and experimental decay constants

Although very high field gradients exist at the tips, the overall response is well modelled by an exponential decay.

In figure 42 the result of a perturbative calculation for the growth of a layer in comparison to the full numerical analysis is shown.

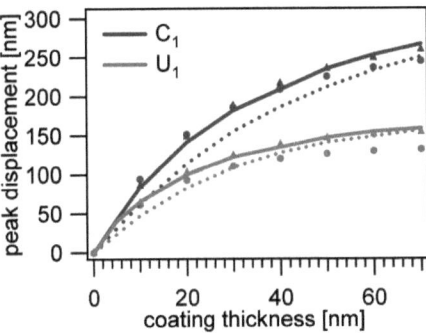

**Figure 42: peak displacement as a function of coating thickness calculated exactly and by perturbation theory. The dashed line is a FEM calculation while the continuous line is a FDTD calculation. The round markers are the perturbative results in FEM while the triangles are from FDTD with correct treatment of the boundary**

Good agreement is found between the descriptions based on FEM, FDTD and perturbation theory. A practical problem however are the errors introduced by the imperfect boundary in FDTD. These errors led to fields at the boundary which seriously distort the evaluation of the perturbation integrals (figure 43a)). One can see the apparent shifts going up again when the perturbation extends out of the near field of the resonator and generally too high shifts. The theory of Lai (equation (55)) incorporates a field at the boundary correctly and there are no problems (figure 43b)). While this does not contradict the developed theory it shows that in practical applications there can be additional sources of error and the perturbative results have to be thoroughly checked. This problem exists only in 3d calculations because only computational resources limit the accuracy of the obtained simulations and demand a compromise between good PML quality and simulation speed, and if the where no numerical errors the problem would not exist.

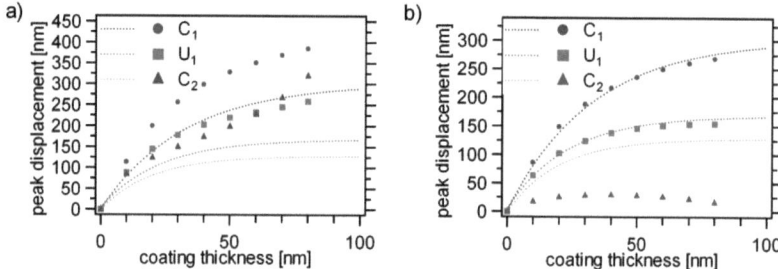

Figure 43: peak displacement as a function of coating thickness calculated by perturbation theory a) quasi-static b) theory of Lai. The dashed lines correspond to the exponential fit to the shifts calculated with FDTD.

### 5.6.3. Response to attached colloids

If single-molecule sensing is to be pursued, one has to consider the response to the attachment of individual objects rather than coating with films or changes in the overall dielectric surrounding. Recently it has become clear that this case must be considered when single molecules should be detected. Only few publications [52] in this field realized the additional effects which are introduced when the detection volume is much smaller than the possible sensing volume of a resonator. In this case the sensitivity becomes a function of the position of the analyte relative to the resonator. Because of this effect it is difficult to calculate the whole parameter space and the following calculations are only examples. A sphere with a diameter of 60 nm and a refractive index of $n = 1.5$ is chosen, which is small compared to the resonator dimensions but still large compared to molecules. This size is suited for testing of the theoretical results because it is large enough to yield a measurable response and is easy to manipulate in experiment. An experiment done like this and showing the local response of the crescents was published as a part of this work [34] and will be described in chapter 6. For the simulation, the sphere was placed on the substrate and touching the crescent. Because the diameter of the sphere is bigger then the height of the crescent the center of the sphere is displaced 10 nm relative to a plane through the middle of the crescent. The spectral response was calculated as a function of its position along the outer perimeter described by the angle $\varphi_s$, compare figure 44a. Figure 44b shows a typical spectrum of the crescent with and without a sphere directly attached to the tip. A small red shift can be observed which is best represented in the difference spectrum. Figure 44c shows this shift as a function of $\varphi_s$. Two effects can be observed. Firstly the shift is strongest directly at the tip, which was expected because the highest field enhancement occurs there. Secondly the shift shows a different dependence on the position of the sphere for each resonance. The physical reason for this effect is directly obvious when comparing this shift with the near field distributions of the three resonances under study which are shown in figure 44d - f. The shift magnitude corresponds to the field strength at the analyte position and in particular no shift is seen if particles are placed in field minima of a certain resonance. Thus multipolar resonances reveal information of the position where a particle is bound. Binding at $\varphi_s = 110°$ for example leads only to shifts of the $C_1$ and $C_2$ resonances while the $U_1$ resonance is unaffected because the field is very low at that position. Similar effects are seen at all positions allowing in principle for a localization of the binding event.

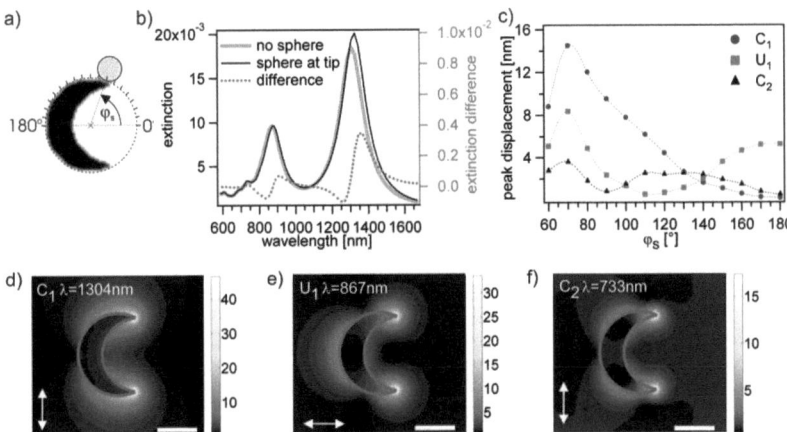

**Figure 44: a) geometry of the crescent with attached colloid b) extinction spectra with and without colloid($\varphi_s = 70°$). c) Response of the crescent as a function of sphere position d)-f): field enhancements ($|E|/|E_{incident}|$) for the resonance peaks at a cut through the middle of the crescent thickness ($z = 20$ nm). Note that the colour scale is not the standard "yellow hot" scale but logarithmically scaled.**

The same analysis was done with perturbation theory and the results are shown figure 45. Good agreement can be seen between the exact and the perturbative results.

The perturbation theory allows now to investigate the shifts as a function of colloid size. This investigation is not possible to perform exactly, as the resulting shifts for small colloids are very small and a practically in 3D simulations not achievable accuracy would be necessary for numerical calculations. The colloid is placed at the position of highest field and touching the crescent resonator. The resulting peak displacement normalized by the change in the dielectric constant is shown in figure 46a.

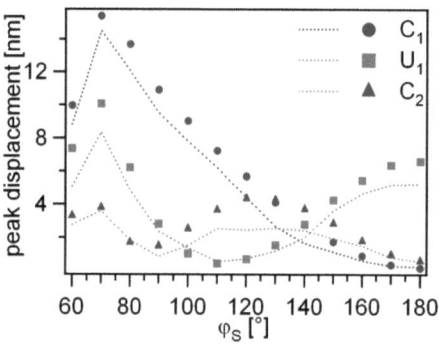

**Figure 45: peak displacement for attachment of a colloid by perturbation theory. The dashed lines correspond to the values calculated with FDTD.**

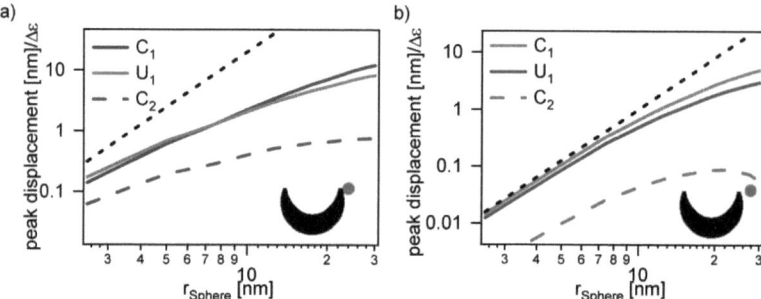

**Figure 46: peak displacement as a function of sphere radius a) at the position of highest field ($\varphi_s = 70°$ attached to the tip) b) in a more uniform field ($\varphi_s = 70°$ 12 nm away from tip) the dashed lines correspond to a cubic function. The insets show the position of the sphere. The displacement is normalized to the dielectric.**

At a radius of 2.5 nm still a shift of 0.15 nm for the $C_1$ and $U_1$ resonance is observed, a value which is reported to be resolvable by state of the art instruments [24]. This size corresponds to the size of a single macromolecule, I.e. streptavidin which is frequently used to evaluate sensing schemes can be described approximately as a sphere with a diameter of d = 5 nm and a refractive index of n = 1.52 [24, 53]. Of course one has to take into account that the dielectric contrast is 5 times lower in aqueous solution which lowers the response also 5 times. Still it is expected that further optimization should allow for detection of single attached molecules given a high enough dielectric contrast. By comparing the slope of the peak displacement with that of a cubic polynomial it is seen that even the smallest modelled spheres do not experience a nearly uniform field where a slope proportional to $r^3$ is expected. When moving the sphere 12 nm away from the tip (at the same angular position $\varphi_s$) into a more uniform field, the slope approaches the expected cubic polynomial (figure 46b) but is of course much smaller. The same should also happen for smaller spheres directly at the tip, but this cannot be tested because of the finite discretisation of the simulation.

## 5.7. Scaling of the sensing properties

In this chapter the effects of decreasing the crescent size is studied. As pointed out in the previous chapter and in chapter 4.3.5 when the size of a resonator gets very small one should approach the quasi-static case where the resonance wavelength is only dependent on the shape of the resonator while the Q factor is only dependent on the materials properties at the resonance wavelength. It has however been shown by Feigenbaum et al. [51], that even for very small resonators retardation effects can play a role and enhance Q. The goal in this chapter is therefore also to investigate weather the sensing properties of the crescent shaped particles can take advantage of such effects.

For this investigation realistically shaped crescent particles are numerically investigated. The particle dimensions according to figure 33 correspond to a particle fabricable in one evaporation step and are the same then in the previous chapter. The outer diameter is varied from 200 nm to 40 nm.

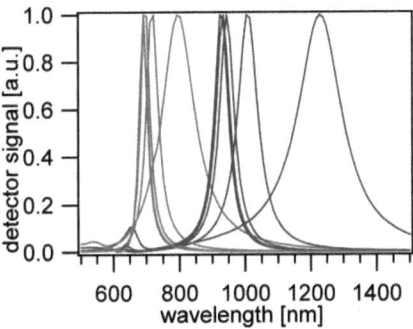

Figure 47: **Shifts of normalized extinction spectra decrasing crescent sizes. The biggest crescent with has a diameter of 200 nm and shows the resonances at the longest wavelengths. The smallest crescent has a diameter of 34 nm. Red lines are U-polarizaion while blue lines are C-polarization.**

Figure 47 shows the resulting normalized far field spectra for this case. It is observed that for decreasing diameter the resonance wavelengths decrease and converge to a final value. Simultaneously the resonances get narrower indicating an increasing Q. For further discussion the resonance wavelengths and Q factors are extracted from the resonance spectra and shown in figure 48.

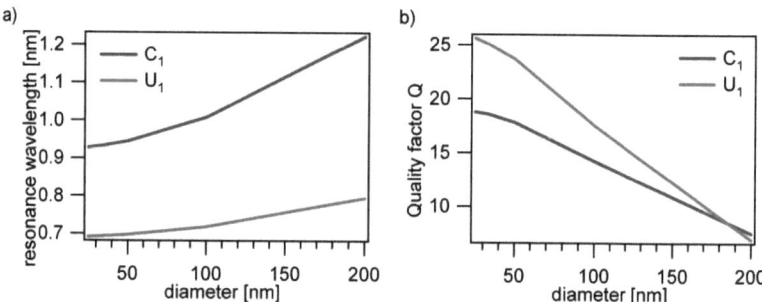

**Figure 48: a) Resonance Wavelength and b) Q as a function of diameter for open crescents**

It can bee seen that while the wavelength converges to a final value, the Q factors increase monotonically. For the $U_1$ resonance the quasi-static Q factor at 700 nm resonance wavelength would be roughly 18 [18]. This value is slightly exceeded at diameters smaller then 100 nm, which means that retardation effects still play a role at small sizes and enhance the Q value. For the $C_1$ resonance the situation is different. The quasi-static value at 900nm wavelength is 18 as well and this value is approached but not exceeded for very small resonators. Contrary to the results from the previous chapter where Q was enhanced trough the interactions of the tips, there is no enhancement of Q for small but open structures. It will now be investigated if closing a small resonator leads to an additional increase in Q.

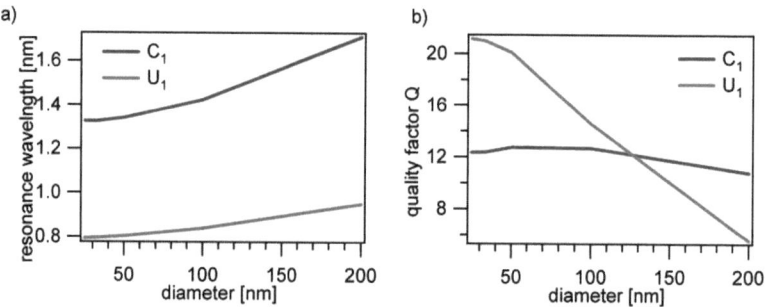

**Figure 49: a)Resonance Wavelength and b) Q as a function of diameter for almost closed crescents**

In figure 49 the same results for an almost closed crescent shaped particle (double evaporation with 120°) is shown. As shown before for the bend rod the resonance wavelength of the second order resonance does not change much compared to the open crescent and converges again to roughly 800nm while the first order is affected by the much smaller gap between the tips and converges to a longer wavelength of 1300 nm. Consequently its quasi-static Q value is decreased to 12 and this value is approached when the particle gets smaller. Between 50 and 100nm diameter Q is slightly larger. For the second order resonance the quasi-static Q is also not exceeded. It can be concluded from this simulations that the quasi-static argument that the Q is only dependent on the resonance wavelength for small particles holds for the crescent shaped particles almost and only small deviations are observed. Q goes down due to radiation when the particles get bigger. Thus it seems that Q can be optimized only by decreasing the particles size, which is of course limited by technical issues to particles bigger than 80nm in diameter (with template sphere lithography, other fabrication techniques might perform better). The next quantity to consider is now the confinement factor C.

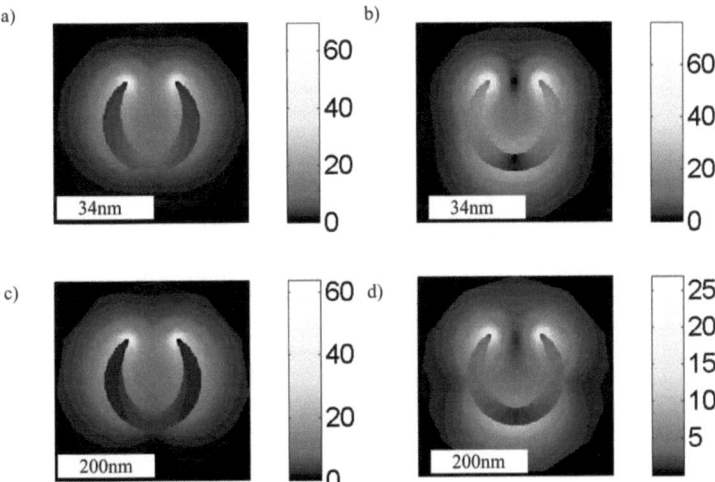

**Figure 50: Field enhancement at a cut trough the middle of the particle in z direction for different sizes and closed crescents. a,c) $C_1$ resonance b,d) $U_1$ resonance.**

Figure 50 and 51 show the electrical field arround crescent shaped particles of different opening angles and sizes. It can bee seen from the plots that the shape and relative field decay remains the same and the plots would not be distinguishable if there would not be the size written at them, except for the different absolute field enhancements which is because of different Q values compare figure 48 and 49. Thus the field confinement and the modal volume scale directly with the size of the particle as was expected from quasi-static theory where the problem is scale invariant. A diameter of 200 nm is already much larger then the quasi-static case and still the scaling holds. [54] This means that the resonance shift should almost scale like predicted for the quasi-static case (see chapter 4.7). The peak shift for attachment of a dielectric colloid as a function of crescent size is shown in figure 52 for different sizes of colloids. For very small colloids the peak shift should scale with $d^{-3}$. This is almost the case for very small colloids as shown in figure 52 as long as the analyte is much smaller then the resonator. For smaller resonators in figure 52a) and larger colloids in figure 52b) saturation of the wavelengths shifts is observed. The analyte extends then in regions of lower field enhancement and decreasing the resonator size does lead then to a much lower shift than expected from the scaling considerations.

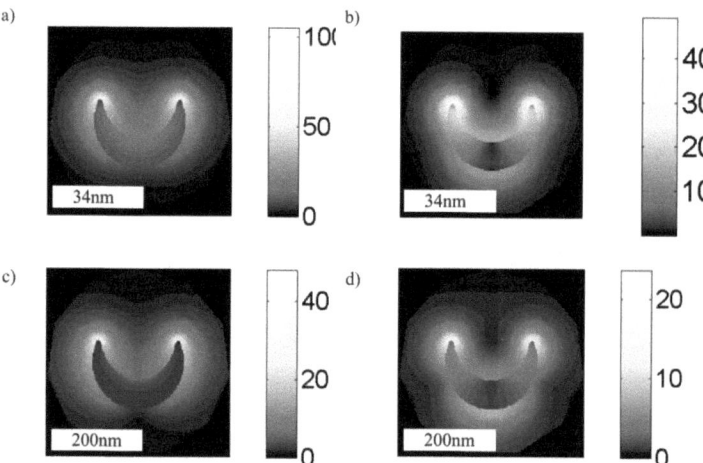

Figure 51: Field enhancement at a cut trough the middle of the particle in z direction for different sizes and open crescents. a,c) $C_1$ resonance b,d) $U_1$ resonance.

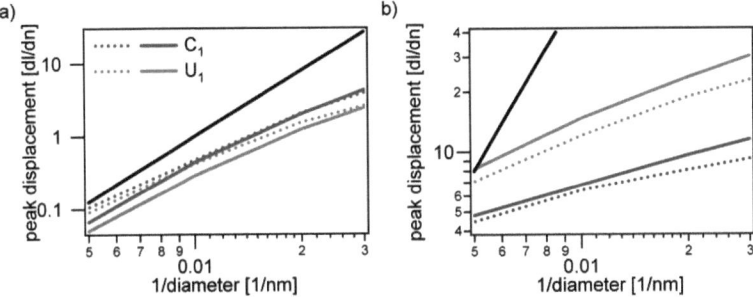

Figure 52: relative peak shift for attachment of a dielectric colloid as a function of inverse crescent diameter for open crescents (continuous lines) and closed crescents (dotted lines). a) small colloid with a diameter of 5nm b) bigger colloid with a diameter of 40nm. The black line is a cubic function which is the expected scaling behaviour for colloids much smaller then the resonator.

## 5.8. Summary and Conclusions

In this chapter the sensing properties of crescent shaped resonators was investigated. A simple and analytical solvable model was introduced in which the observed resonances were attributed to standing waves along a metal rod. In this model the existence and excitation conditions for the resonances could be well explained. Also the sensitivity of the crescent shaped resonators to changes in the surrounding refractive index could be qualitatively explained in this model. 3D numerical simulations where then carried out to achieve a more quantitative description. A full characterization including the sensing properties for a realistic crescent shaped resonator was carried out. The obtained results for the sensitivity were compared with perturbation theory from chapter 4 and good agreement was found. Finally the scaling of the crescent sensitivity with geometry parameters and size was studied.

It was shown that even for resonators larger than the quasi-static limit the field confinement factor scales as expected for the quasi-static regime. Due to the sharp tips strong confinement for localized changes in refractive index like single molecules can be achieved. The Q factors of this resonators achieve the quasi-static values at diameters around 50nm. This is too small to fabricate with the current state of the art. At fabricable diameters of 100nm however still Q factors of 15 to 20 are achieved. At this sizes peak shifts of 0.5nm/RIU can be achieved for molecules binding directly at the tip, which would correspond to shifts of 0.1 nm for the dielectric contrast of a single streptavidin molecule in water. The detectability of such an event is strongly dependent on the SNR on the actual experimental case as described in chapter 2. From equation (48) the SNR needed to resolve such an event can be estimated. With the line width of 50nm of the $U_1$ resonance a shift of 0.1nm and f=0.5 a SNR of 250 would be needed to be able detect the binding of a single molecule. These predictions will be checked experimentally in the next chapter for an easy to fabricate and well defined model system.

Another interesting result of this chapter is that the different resonances with inhomogeneous field distributions will lead to different sensitivities as a function of the binding position of a molecule on the resonator. The existence of multiple resonances can therefore be used to determine the position of a binding event or when a layer grows on the surface of a resonator to determine multiple parameters of the layer like thickness and refractive index independently.

The results of this chapter enable the reader to quickly select the optimum geometry of a crescent resonator for a particular sensor design.

# 6. Local response experiment

In this chapter experiments to show the response of plasmon resonances to local changes in refractive index are shown. In this experiments polystyrene colloids of around 60nm diameter where used to change the refractive index only at the tip of a crescent shaped particle. While this size is still large compared to a molecule this approach has the advantage of being easy to handle in an experiment. The colloids can be easily manipulated with an scanning force microscope (SFM) and additional errors from working in aqueous solutions like unspecific binding or changes in refractive index of the liquid are avoided. This kind of experiment is therefore suited to test the theory developed in chapter 4. From the results of this chapter together with the theory an extrapolation can be done to estimate the sensitivity limits of the crescent shaped particles. The results of this chapter where published already [34].

In order to estimate the local field enhancements and resulting sensitive positions, the optical response of the crescent was modelled by a FDTD code, see chapter 3.2. A simplified crescent shape in three dimensions with a top and bottom flat plane and vertical side walls was assumed. A clear resonance around $\lambda$ = 880 nm is seen in the scattering cross section shown in figure 53a). The field enhancement at resonance ($\lambda$ = 882 nm) is shown in 53b). A significantly enhanced electrical field is seen which is mainly confined to points with a distance of less than 30 nm from the crescent surface. Three regions of enhanced field, two at the tips and one near the middle of the crescent are seen, separated by planes with low field. To get some intuition for the field confinement of this structure, the modal volume of this resonance as the volume around the tips of the resonator where the field exceeds 10% of its maximum value is calculated. This volume has an approximate value of

$10^5$ nm$^3$, three orders of magnitude smaller than estimated for a focus of a microscope objective with a numerical aperture of 1.4 at a wavelength of 800 nm. The strongest field enhancement occurs at the tips where a typical decay length for the field to exp(-1) of it's maximum value which is of the order of 20 nm.

The resonance shift upon attachment of an analyte is connected to the electromagnetic energy density of the plasmonic resonance as was described in chapter 4.3. In a first order approximation, the spectral shift is proportional to the fraction of the energy within the volume occupied by the analyte. This implies that, in order to obtain a large shift a small total modal volume is advantageous and the analyte should be placed such that it experiences maximum field. Therefore, a colloid position next to one of the tips as indicated in 53b) is choosen. In the numerical model, this leads to a resonance shift of 8.5 nm, big enough to be measured in an experiment.

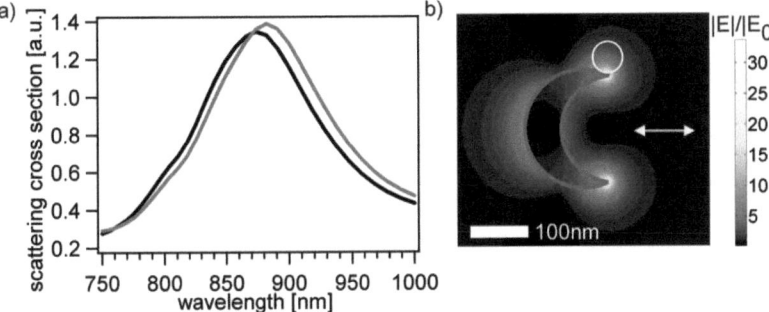

Figure 53: a) Calculated scattering spectra with (red) and without colloid (black). b) calculated magnitude of the electrical field at a plane trough the middle of the crescent. The position of the model analyte for the shift shown in a) is indicated schematically. The arrow shows the incident polarization

## 6.1. Fabrication and Experiment

### 6.1.1. Fabrication of the crescent samples

Gold nanocrescents with a diameter of 180 nm to 200 nm were prepared on a microscope coverslide by nanosphere template lithography [13, 14]. Polystyrene masking colloids with a nominal diameter of 150 nm were dispersed on the substrate and a gold film evaporated at an oblique angle. Etching with a normally incident ion beam leaves only the crescent-shaped gold structures in the geometrical shadow of the masking colloids. Finally the masking colloids were removed. After production of the crescents a second gold evaporation through a grid as used for transmission electron microscopy ( TEM grid 75 µm x 300 µm ) was carried out, resulting in a structure of rectangular gold pads of 80 nm height. These pads were used as guiding structure to be able to uniquely identify the crescents under consideration in the optical microscope, the scanning electron microcope and the atomic force microscope. After fabrication of the guiding structure the sample was immersed in toluene for 24 hours to ensure removal of residues of the masking colloid. Figure 54 shows the guiding structure obtained by evaporation through an TEM grid. With a syringe needle, an additional cross was scratched in the surface, leading to partly removed and cut gold pads as seen in figure S1 a). These structures were then used as guiding structures to identify identical areas in scanning electron microscopy (b) and optical microscopy (c). Then Polystyrene (PS) colloids with a nominal diameter of 60 nm were randomly distributed on the surface. 10 µl suspension of polystyrene colloids (diameter 60 nm, Polystyrene Nanobead: NIST, Polysciences, Warrington, USA), diluted 1:100 in ethanol was dropped onto the sample. After 10 s the ethanol was blown away with nitrogen, leaving a dilute coverage of the sample surface with polystyrene colloids.

### 6.1.2. Determination of the optical response

The optical response of an individual crescent to the attachment of a single colloid was determined as follows. First, single object scattering spectra of 20 randomly chosen crescents were recorded. One of this structures had a colloid dimer attached and was not considered in further analysis. A subset of 9 crescents was chosen to which PS colloids were pushed by a scanning force micropscope (SFM). Manipulation of the colloids was done with a scanning force microscope (SFM, Nanoscope IV, D3100 closed loop, Veeco), used as nanomanipulation tool (nanoman, Version 6.11+1). First a scan of the area of interest was carried out in non-contact mode. Hereby, the crescents under consideration were identified as well as nearby colloids. Then, the SFM tip was positioned next to a PS colloid and contact mode operation was enabled. The feed-back was switched off and the tip-sample distance decreased by 10 nm. By moving the tip along a line across the PS colloid, the latter was pushed in the direction of the tip movement (1 µm s$^{-1}$). This procedure was repeated until the PS colloid reached the desired position and the resulting structure was finally imaged again in non-contact mode. In figure 55a) the nanomanipulation is illustrated and in figure 55b,c) the sample topography before and after manipulation is shown. The remaining 10 crescents served as control where no PS colloid was pushed. Then optical spectra all colloid were recorded again. Finally, the combined crescent-colloid resonators were examined by scanning electron microscopy (LEO 1530 Gemini). The center wavelengths of the resonance peaks were determined by fitting a Lorentzian function to the spectrum. The spectral shift was determined as the difference between the peak positions before and after approaching a colloid.

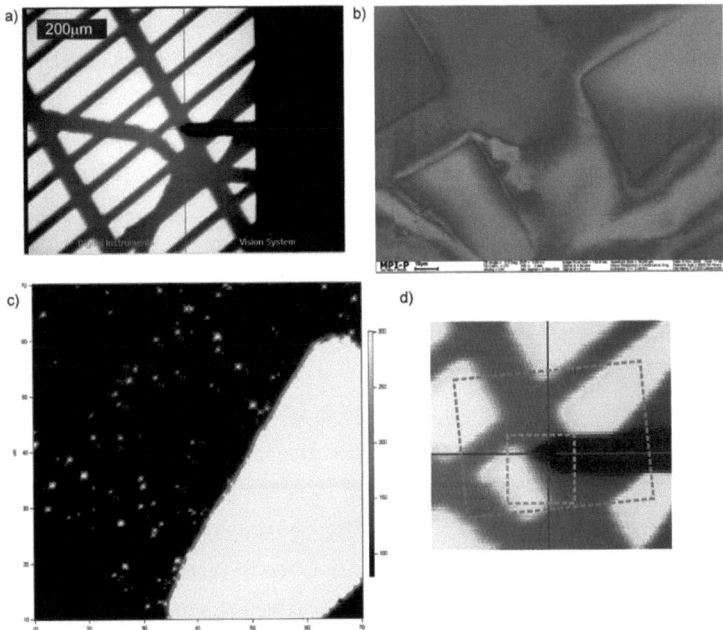

**Figure 54: Guiding Structure a)** as seen in the light microscope which is part of the SFM. The cantilever is seen approximately in the middle of the image. **b)** as seen in the SEM **c)** as seen in the scanning confocal microscope. **d)** close up of a) with the position of the SEM (red) and confocal (blue) microscope images indicated as dashed rectangles. Note that the confocal image is mirrored at the vertical axis.

### 6.1.3. Measurement of the spectra

Single particle spectra were measured with a homebuilt inverted confocal dark field scanning microscope. Light from a xenon lamp (Mueller Electronic-Optic, Germany with Osram XBO150) is coupled into a single mode fiber. After the fiber, the beam is collimated with a 150 mm focal length achromatic lens. The inner part of the 23 mm diameter beam is blanked by a circular disc of 5 mm diameter for dark field high NA illumination. The beam passes a broadband 50/50 beam splitter plate and is focused on the sample by a microscope objective (Nikon Plan-Apo 60x, NA = 1.4). The sample is scanned by a XYZ piezoelectric stage in closed loop mode (Tritor 101 CAP Piezosysteme Jena GmbH, Jena Germany). The back reflected light is collected by the same objective and partly redirected by the beam splitter to the detection channel. There, the specularly reflected light is blocked by a circular pupil of variable diameter which is adjusted to give optimum signal to noise (typically for diameters 5 to 4 mm). After this, the beam is focused to a pinhole of 100 µm by a

70 mm focal length achromatic lens and collimated again by a lens of the same type. A Glan-Thomson polarizer selects the polarization direction which corresponds to the resonance peak under investigation. For imaging, the light is spectrally filtered by a band pass filter with 850 nm center wavelength and 10 nm line width and focused to an avalanche photo diode (Perkin Elmer, Optoelectronics Inc., USA). To record single particle spectra, the light is coupled to a multimode fiber (200 µm core with NA = 0.22) and directed to a grating spectrometer (Andor Shamrock SR303i, Andor Technology, Belfast, Northern Ireland).

A spectrum is recorded in the following way. First an image of the sample is taken by scanning the sample with the piezo stage and recording the scattered intensity from the sample as a function of position. A particle, identified as bright spot, is moved into the focus. Fine tuning by manual offset regulators ensures to have the particle in focus with an accuracy of 20 nm. Then the scattered light is directed to the spectrometer by an inserted mirror and the spectrum is recorded.

For each spectrum $I_C$ of a crescent, a background spectrum $I_B$ very close to the crescent under consideration was recorded. The latter contains only the reflection at the glass surface and stray light from the optical setup. A reference spectrum $I_R$ was recorded at the rough gold surface provided by the guiding structure immediately afterwards. The spectra of $I_C$, $I_B$ and $I_R$ are all recorded within 5 minutes reduces the effect of long-time drift of the spectral response of the microscope. The crescent spectra were calculated as

$$I_{Scat}(\lambda) = \frac{I_C - I_B}{I_R - I_B} \tag{104}$$

No further smoothing algorithms were applied. The spectrum was then normalized to a maximum intensity of one. Peak positions were extracted by fitting a Lorentzian:

$$I_L(\lambda) = \frac{2A}{\pi} \frac{\Gamma}{4(1/\lambda - 1/\lambda_c)^2 + \Gamma^2} \tag{105}$$

to the data. Here A is the area under the peak $\Gamma$ the full width at half maximum (FWHM) and $\lambda_c$ the center frequency of the peak.

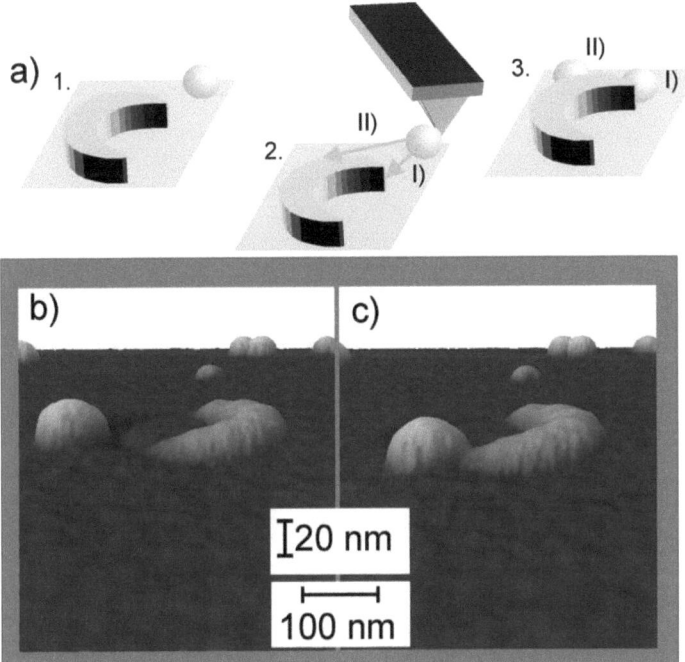

Figure 55: a) Schematic representation of the experiment. 1. Some polystyrene colloids are deposited on a crescent-decorated glass surface. 2. By SFM nanomanipulation a single colloid is approached to the crescent tip. 3. structure after nanomanipulation I) and II) illustrate two possible final colloid positions. b) Topography images of one crescent before manipulation. c) after manipulation.

## 6.2. Results and Discussion

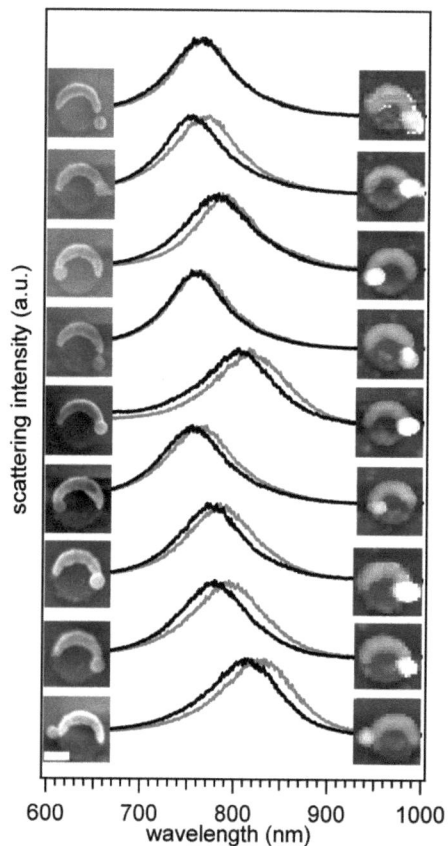

**Figure 56: Spectra of the crescent shaped particles before (black) and after (red) attachment of a PS colloid to the tip for each crescent are shown. For each crescent, electron- (left) and scanning force micrographs (right) after manipulation are shown. The scale bar in the bottom left graph is 100 nm and applies to all images.**

Figure 56 shows scattering spectra for the 9 crescents. In all cases a resonance is seen that can be described to a good approximation as a Lorentzian. After attachment of the dielectric colloid, in 7 cases a clear red shift of the maximum by about 11 nm is seen.

A first observation is that there is no shift in two cases. The reason for this is not apparent from the SFM images shown at the right side of figure 56. They suggest attachment of a colloid for all cases but small gaps cannot be resolved due to the finite size of the SFM tip. An inspection by electron microscopy (left side of figure 56) reveals the reason for the special behaviour of these two crescents: Here, the PS colloid has a distance around 10 nm to 20 nm from the crescent while in all other cases the PS colloid appears to be in direct contact. This observation represents an experimental verification of the predicted very high field localisation at the tip since this small displacement is sufficient to largely decrease the optical effect of the PS colloid. Next, only the crescent response to directly attached particles is considered.

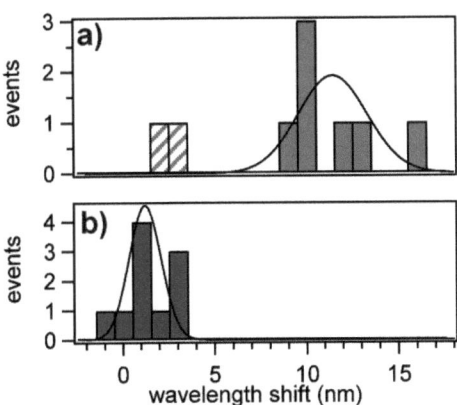

**Figure 57: Histogram of spectral of shifts of particles where a colloid was attached (a) and control particles where no colloid was attached (b). Two particles which have a finite gap between crescent and particle are marked hatched and are not taken into account for the further analysis. The Gaussian curves illustrate the mean value and standard deviation of the distributions.**

Figure 57a) shows a histogram of all measured shifts of modified crescents. For direct contact between crescent and colloid, shifts between 8.8 nm and 16.5 nm with a mean shift of 11.5 nm are seen which is of the same order as the result predicted by the model (8.5 nm). This points to a high quality of the crescents in two respects. Firstly, all shifts are of the same order. This reproducibility for different crescents is important if they are to be used for the analysis of an unknown analyte. Secondly, the measured shift approximately agrees with the calculated shift which points towards a high field concentration at the tip. This agreement between theory and experiment within less than a factor of two is even more remarkable when considering the significant difference between the idealized crescent shape used for the calculation (compare sketch in figure 55)and the reality which is unavoidably imperfect, see e.g. the scanning electron images in figure 56 and the SFM topography shown in figure 55b,c). The near field distribution and in turn the spectral response appears to be tolerant to moderate shape imperfections which is important for possible applications.

Furthermore, the spheres are attached at different positions relative to the tip but all cases lead to similar shifts This insensitivity can be rationalized by the field distribution in 53b). Around the tips an almost spherically symmetric field exists which would lead to the observed behaviour. The experimental accuracy and possible artifacts can be addressed based on the measured wavelength shifts of unmodified crescents (figure 58 shows the spectra)

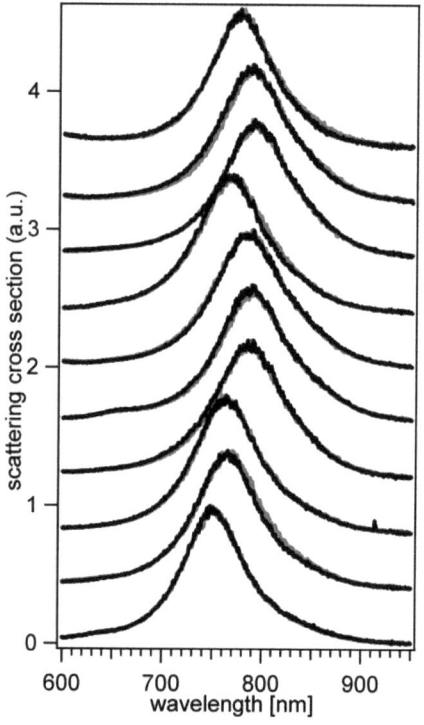

**Figure 58: Spectra of the control structures that were not manipulated**

Their distribution, shown 57b) has a mean value and standard deviation of 1.2 nm ± 1.2 nm, respectively. A shift in the control crescents resonance wavelength could due to contamination or modification by the SFM tip[55] or drifts in the optical setup could be anticipated. The measured distribution suggests a range from 0 nm to 2.5 nm for these effects . Furthermore, the width of the distribution yields a measure for the absolute accuracy of the measured wavelength shift.

All crescents with PS colloid decoration shift slightly more than predicted by the calculation, which assumes a highly simplified geometry. Perfect agreement with the experiment, both in terms of a resonance wavelength at 770 nm and in a shift around 11 nm is obtained if a small glass post below the crescent which is due to the preparation process is included in the simulation. An interpretation of these small improvements in the matching of theory and experiment should take into account other mechanisms which may lead to effects of the same order. Possible additional mechanisms that should be considered are water menisci [55, 56] or impurities near the contact point of PS colloid and crescent.

Next, it is discussed how the individual differences between crescents can be understood, based on this data. From the optical spectra and electron micrographs four parameters describing each crescent can be found: The resonance wavelength, the resonance shift and the sizes of the colloid and the crescent. Furthermore, the position of the colloid varies but this parameter is not straightforward to quantify.

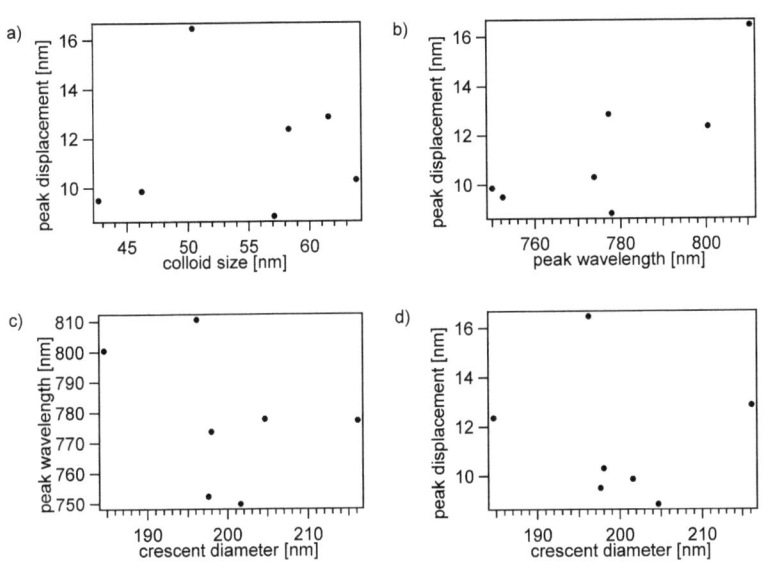

**Figure 59: Correlations between the parameters of the crescent and the peak displacement**

No strong correlations between either of these parameters are found. In particular, the small variations in resonance wavelength and peak displacement are independent from the small size variations of the crescents. Therefore, microscopic differences between the crescents, i.e. different grain boundaries, small cracks or protrusions or variations of the glass support have a stronger influence on the resonance than the small differences in crescent size. Significantly different colloid diameters between 47 nm and 63 nm can be obtained from the electron micrographs. Although the colloid volumes differ by more than a factor of two, this quantity does not show a clear correlation to the observed shifts. Since a shift proportional to the volume would be expected for colloids in an approximately homogenous field, this observation points again to a strong field localization at the crescent. Mainly the part of the colloid next to the contact area experiences a high field and contributes to the shift. A plot of the observed shifts vs. the resonance position (see supporting information) shows a slight positive correlation, which points towards an increased shift for long wavelength resonances in agreement with the theory from chapter 4.3. The largest shift is found for the only sample where the colloid seems to be placed next to the crescent (lowest curve in figure 59) rather than on top as seen for all the other cases. This shift differs by 5 nm from the mean value, corresponding to four times the statistical uncertainty. From this data it is suggested that side attachment and long-wavelength resonance, may both lead to an increased shift due to physical mechanisms. Since the geometrical reasons for the variations in resonance wavelength are not clear, it is not possible to model this effect. To clarify the different behaviour for colloids at the tip and at the side of the crescent additional simulations were carried out. Three different configurations according to figure 60 were considered. Also a more realistic crescent tip was extracted from the scanning force images, allowing for a configuration where the crescent tip extends to positions slightly below the analyte as observed in the experiment. One configuration was simulated assuming a glass post below the crescent due to a slightly too long ion etching in the production process.

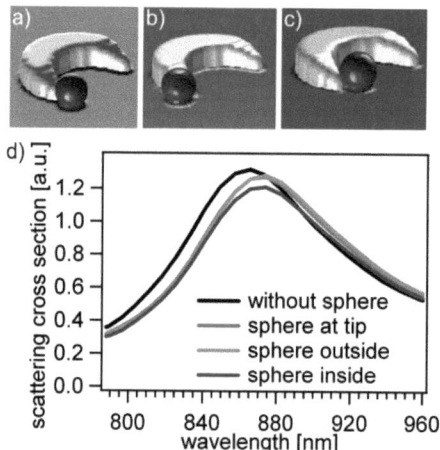

**Figure 60: simulated geometries of crescents with more realistic rounded tips and colloids attached at different positions. Attachment to the tip (a), to the outer side (b) and to the inner side (c) are considered. The spectra are shown in d). Shifts of 7.5 nm for (a), 8.5 nm for (b) and 7.4 nm for (c) are found.**

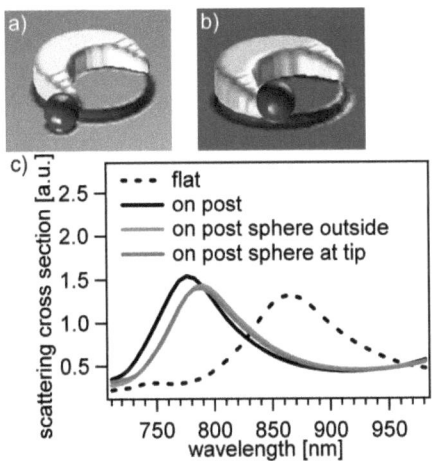

**Figure 61:** a, b) Sketches for two configurations which take into account a glass post of 20 nm in height below the crescent. c) resonances for this configuration. the resonance of the bare crescent without analyte is shown as dashed line for comparison. Shifts of 12 nm and 10.3 nm for tip- and side-attached analytes are found.

Figure 61 shows the response to a side attached colloid for an overetched crescent on a 20 nm high post. For this geometry a blue shift of the spectrum as well as a higher sensitivity to the attachment of a colloid can be seen.

It can be summarized that attachment of the colloid at the crescent tip leads to a reproducible spectral shift which is in agreement with the theoretical prediction and tolerant to slight variations of the sensor and the analyte.

## 6.3. Extrapolation to Single molecules

In the following, it is discussed what conclusions can be drawn from these experiments for label-free biosensing. There, significantly lower shifts are expected for two reasons. Firstly, the model analytes used for this study had a diameter of $d_{PS} = 60$ nm which is significantly larger than typical analytes in biological experiments. I.e. streptavidin which is frequently used to evaluate sensing schemes [24, 53] can be described approximately as a sphere with a diameter of $d_{Strep} = 5$ nm. This corresponds roughly to a 100 times smaller shift. Furthermore, the spectral shift is approximately linear in the refractive index difference between analyte and surrounding which is $n_{PS}-n_{air} = 1.58-1 = 0.58$ for the model discussed here and $n_{protein}-n_{water} \approx 1.52-1.33 = 0.19$ for streptavidin in water. This corresponds to another reduction by a factor of 3 in expected wavelength shift. Together, an

300 times smaller effect for binding of a single streptavidin molecule is expected, corresponding to a wavelength shift of 0.04 nm.

This is far beyond the experimental accuracy of 1.2 nm that we achieve here. However this deviation is largely dominated by systematic errors. The sample was moved from the optical microscope to the SFM and back, so both contaminations and accuracy of optical alignment are expected to play a role. These problems would be avoided or at least significantly reduced in an in-situ experiment where the spontaneous binding and unbinding of an analyte is detected.

From the data a coarse extrapolation to the sensitivity for smaller objects and in different environments can be done, provided systematic errors are eliminated. Then, the signal-to-noise ratio limits the accuracy in the determination of the maximum wavelength. The data was recorded with a signal-to-noise ratio of 70. This leads to a statistical precision of 0.3 nm for the determination of the peak positions which is approximately eight times the expected shift for a single molecule. Several straightforward experimental improvements promise to bridge this remaining gap to real label-free single biomolecule detection. The integration time can be increased for a better signal-to-noise ratio but the accompanying loss of time resolution limits the usability of this strategy. Therefore, the signal strength for fixed integration time must be increased. While the detection efficiency is already quite good, stronger light sources for illumination could be considered. Here, the tolerable degree of sample heating represents an upper limit. An additional improvement can be achieved by restricting the illumination spectrum to few wavelengths which are chosen to minimize the uncertainty in peak position fitting [57]. Independently from these improvements of the experimental setup, optimization of the resonator shape may be alone sufficient to increase the experimental precision to single-molecule sensitivity.

In conclusion, a method was presented to measure the response of the plasmonic resonance of crescent shaped metal nanoparticles to the attachment of a dielectric colloid. Both the spectral shift and the analyte position were determined, allowing for a quantitative comparison of the experimentally measured spectral shifts with the theoretical prediction. Very good agreement with a model calculation was found which assumes bulk dielectric response for the metal and an ideal geometry. In particular, a strong field confinement at the tips of the structure is shown by our data. Extrapolation to smaller analytes on the molecular scale encourages further efforts towards label-free single biomolecule detection which will probably be achieved in the near future.

# 7. Dense Arrays of crescent shaped resonators

In the last chapters single crescent shaped particles were characterized and their sensing capabilities were shown. Advances in fabrication techniques allow now also to create dense ordered layers of these particles [58]. The question now arises if dense, ordered layers show additional effects due to the interaction between the single particles. Actually these structures are very similar to the so called metamaterials which gained a lot of research interest in the last years due to their promise of providing new otherwise unachievable optical properties like magnetism in the optical regime or lefthandedness which could lead to new imaging devices that beat the diffraction limit or provide optical cloaking. The chapter is organized in the following way: First an outline of possible coupling effects is given and the available literature is reviewed. Then the setup of calculations is described and numerical results of coupling effects are presented. Finally the numerical results are compared to experimental data obtained from samples produced by colloidal lithography with dense monolayers of colloids.

## 7.1. Coupling Effects

A resonator arranged in an array will experience and additional field which comes from the scattered field of its neighbours. These contributions will sum up coherently at the resonators place and give rise to interference effects. Depending on the phase and therefore distance between the particles this can lead to constructive interference where the resulting field will drive the particle in phase with the incoming field or destructive interference where incoming and scattered field are out of phase by 180 degrees. The former happens when the distance between neighbouring particles is an integer multiple of the excitation wavelength and leads to enhanced absorption and sharp additional features, so called Woods anomalies, termed after their first observation by Wood in 1902 on two-dimensional gratings [59]. For two-dimensional gratings, they have been discussed in the context of the surface-enhanced Raman effect[60]. The theoretical concepts of Wood anomalies in two-dimensional gratings have been reviewed recently by de Abajo [61]. Numerical simulations of specific grating geometries are available [62, 63] as well as experiments for a variety of lattice geometries and types of plasmonic resonators [7, 64-67]. Generally, Whenever a diffracted order becomes evanescent or matches a surface mode, an anomaly can occur which may be much sharper than the resonance of the isolated metal particle. The particle resonance starting from far separated particles first red-shifts with decreasing particle spacing. Whenever a diffracted order becomes evanescent, the shift direction reverses and turns into a blue shift with decreasing particle distance. This blue-shift is observed as long as the individual particles are still sufficiently far apart that no strong near-field interaction occurs.

When the particles are closer together, so that their size is not negligible compared to their distance this model breaks down because the particles do not interact like point dipoles anymore. In principle one could consider the same model and take the radiation of higher order multipoles into account but this approach get tedious very fast and doesn't provide more inside than full numerical calculations. When the particles are very close packed and their near fields overlap even this method is not practicable anymore. When the interaction between particle is entirely dominated by the near field in a first approximation one would expect that retardation plays no role and the near field is in phase with the excitation field. This would lead to a stronger driven resonance since the local electric field at the resonator is enhanced and hence absorption and emission are enhanced which leads to a lowered Q.

Practically distances below half the wavelength are the most important for two reasons. The first is that in order to get big effects from a layer of resonators they should be as dense as possible. Second in this regime no diffraction orders beside the zeroth order exist and the layer behaves like an anisotropic homogeneous layer when observed in the farfield. Coupling between resonators is dominated by the near field. Given a crescent shaped resonator with it's resonances above ca. 800 nm and resonator sizes of 100 nm to 200 nm, an interparticle spacing between 150 nm and 500 nm should be studied. This distance can be further split in two regions. A weak coupling regime where the near fields of adjacent resonators only overlap very little and do not change shape much and a strong coupling regime where the resonators are closer together than the penetration depth and the near fields and properties change qualitatively.

## 7.2. Setup of the simulations

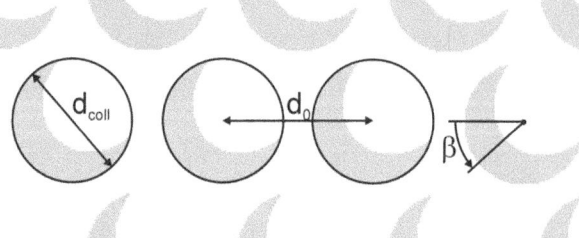

Figure 62: Geometry considered for the calculations. Crescent contours as used for the calculations are arranged in a hexagonal lattice as sketched in grey. The diameter of the masking colloid $d_{coll}$, the nearest-neighbor distance $d_0$ and the angle $\beta$ between the crescent symmetry plane and the axis connecting nearest neighbors is indicated. Due to the six fold symmetry of the underlying lattice, the angle beta is restricted to $0° < \beta < 60°$.

Figure 62 shows the geometry under consideration. Assuming round masking colloids, and an idealized lithographic process the crescent structures which are obtained by colloidal lithography can be described as a sphere cut by an ellipse, leading to the ‚crescent shape' shown and described in chapter 5.2.

For the calculations, metal structures with vertical sidewalls and a thickness t=40 nm are considered. The dielectric function of gold as tabulated by Johnson and Christy [20] is used and the crescents are supported by glass (refractive index n=1.5) and extend into air. Here, crescents in hexagonal arrays are investigated. To fully describe this geometry, in addition to the diameter $d_{coll}$ of the masking colloid, the nearest-neighbour distance $d_0$ and the angle β between the crescent symmetry plane and the hexagonal grating must be specified. For the simulations the FDTD method was used. Simulations where done as described in chapter 3.2 with the exception of periodic boundaries used here.

## 7.3. Effect of crescents distance

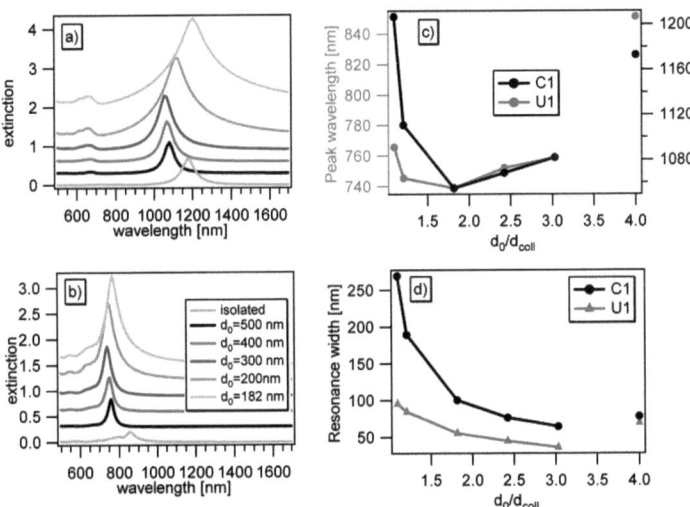

Figure 63: a,b) Extinction spectra ($d_{coll}$=165 nm, t=40 nm, β=45°) for different nearest-neighbor distances $d_0$ for U and C polarisation The spectra are vertically offset for clarity. c) peak wavelength extracted from a,b. d) resonance width extracted from a Lorentzian fit to the spectra shown in a,b.

First, by numerical calculations on isolated crescents as well as hexagonal crescent arrays the effects of crescent coupling in periodic arrays is investigated. Extinction spectra for U and C polarised light are shown in figure 63 a and b respectively. The calculations stay in the sub-wavelength regime and no other than the zeroth diffraction order exists.

Peak positions for the $U_1$ and $C_1$ resonances as extracted from these calculations are shown in c. For comparison also the peak positions from isolated crescents are shown. In both cases, the resonance of a loose array ($d_0/d_{coll} = 3$) is blue-shifted significantly from the resonance of the isolated crescent. With decreasing particle distance the resonances first shift towards the blue before, at very small resonator separations ($d_0/d_{coll} < 1.5$), a red shift occurs. The blue-shift marks the weak coupling regime and is in agreement with earlier observations on arrays of weakly coupled plasmonic resonators [7, 67, 68]. This blueshift can be understood from first order perturbation theory [32]. If the dielectric constant $\varepsilon$ within the mode volume of a resonance changes by $\Delta\varepsilon$, this induces a shift which is proportional to $\Delta\varepsilon$. Since $\varepsilon$ is negative for metals a blueshift results. The reverse shift at even smaller separations indicates that the perturbative approximation breaks down and strong coupling between the resonances sets in. The transition from weak to strong coupling occurs al larger distances for C1 as expected from the larger modal volume of this resonance.

The resonance width is diplayed in d) Arrays with ($d_0/d_{coll} = 3$) show a narrower resonance than isolated resonators, this effect being most pronounced for U1. This may be related to the different scattering channels available for the two cases: isolated crescents can scatter in all directions while a zero order grating only redistributes energy between the transmitted and reflected plane wave. With decreasing $d_0$ the strong coupling regime is reached and the resonance width increases significantly.

The maximum absorption increases with decreasing d0 as is evident from a,b, in agreement with the expectation for a more and more close-packed arrangement. The extinction crossection is in general significantly higher than the geometrical crossection of the metal structures. For $d_0$=200 nm a maximum extinction of more than 2 is reached, corresponding to only 1% of the incident light being transmitted while the metal coverage is only 28% of the unit cell for this geometry.

## 7.4. Effect of crescents rotation

As an additional degree of freedom, the angle of rotation $\beta$ of the crescent relative to the underlying hexagonal lattice, compare figure 62 should be considered. Figure 64 a and b show the effect of varying $\beta$ for the closest packing considered ($d_0/d_{coll} = 1.15$). Some variation in peak position is seen but the effect is small in spite of considerable differences in the near field distribution for $\beta =0°$ and β =30° for both U1 and C1 which are displayed in Figure 62a). The variations in extinction spectra are smaller than typical sample-to-sample variations found in experiments.

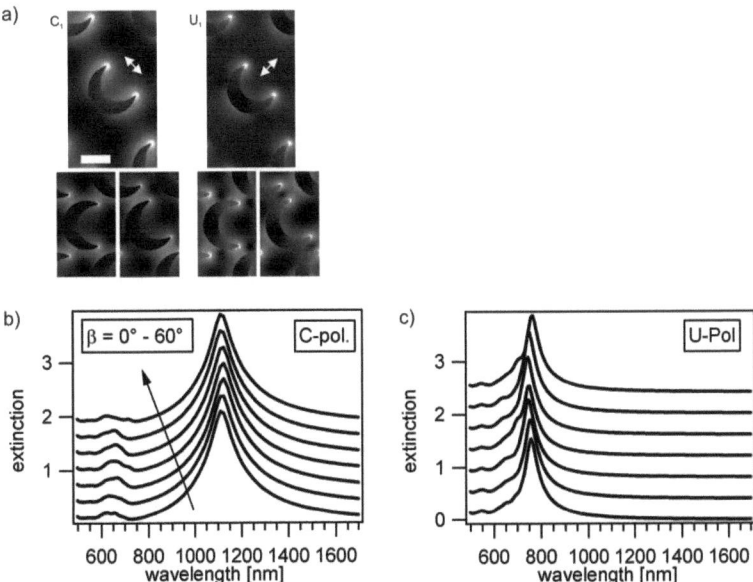

**Figure 64: a) Field distribution at resonance for d0=500 nm and d0=185 nm for 0° and 30° relative orientations between the crescents and the underlying grating. b,c) Extinction spectra for varying rotation angles and the two polarizations.**

## 7.5. Effect of crescents deformation

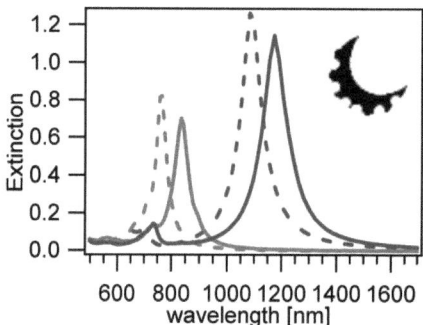

**Figure 65: Deformation of the contour. The dashed lines show the spectra of the undeformed crescent.**

The true shape of metal nanostructures often deviates significantly from the ideal geometry. To get some feeling for the impact of such variation on the optical response, crescents with a distorted shape similar to real shapes (from SEM images) were modelled. Their optical response is shown in figure 65. For these highly deformed shapes strong spectral changes are seen, resonance peaks shift by more than their width. Because it cannot be expected that all rough particles behave the same way this will result in an inhomogeneous broadening and therefore also an extinction reduction.

## 7.6. Experimental results and discussion

In this chapter the calculated response is compared to experimental results. This work was done in collaboration with M.Tamm, N.Vogel and N.Bocchio who prepared the samples and did the optical characterization. The methods used to produce the dense arrays of particles are described in [58].

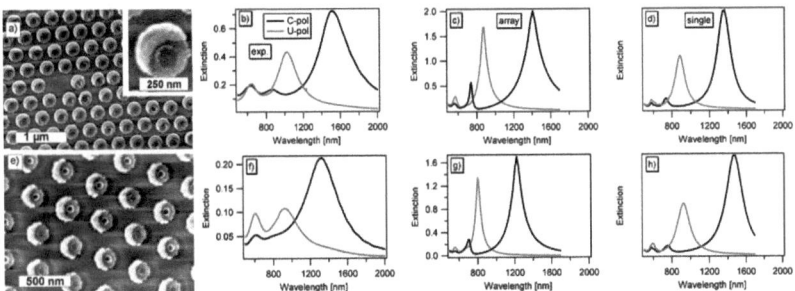

**Figure 66: Experimental results in the weak coupling regime. a) ,b) and e),f) are experimental results while c) and d) are the corresponding calculations. e) and h) are corresponding spectra of single isolated crescents.**

Figure 66 shows crescent arrays with distances which correspond to the weak coupling regime discussed above. Calculations of perfect arrays with the corresponding values for $d_0$ and $d_{coll}$ as well as calculations for single crescents are shown for comparison. In the upper row, the crescent shape appears quite 'clean' in the electron micrograph. Some defects are present in the lattice. The measured extinction corresponds to a fraction of 20% and 40% of transmitted light for $C_1$ and $U_1$ respectively. These values are larger than the crescents geometrical cross section (20%) but much smaller than the values predicted by the calculation. For the high extinctions predicted, already small imperfections which transmit some light lead to a strong reduction of the measured extinction. Small differences between individual resonators as well as defects in the lattice may both account for this effect. The theoretically predicted narrowing of U1 is not seen in the experiment, but a maximum which is broader than predicted both for isolated crescents and for arrays. This, in contrast to the very good agreement between theory and experiment for isolated crescents (see figure 40) points to a high sensitivity of the optical response of crescent arrays to fabrication imperfections. To get an idea how much of the effect is caused by lattice defects like grain boundaries or voids and how much by other effects like inhomogeneous broadening one could calculate how much lattice defects one needs to explain the differenced between calculations and experiment in figure 66. For sample a) this gives 20% voids which is reasonable. The broader lines observed in experiment however show that to some extend also inhomogeneous broadening plays a role. Figure 66e-h) (lower row) shows a sample with a smaller d0=220 nm as obtained by longer etching. Now, the shapes are clearly distorted which explains why the experimental extinction is strongly reduced relative to the theoretical prediction. Due to the strong size reduction, the masks are attain non-perfect shapes which are then inherited by the crescents. Again for sample e) if the extinction reduction would be only caused by lattice defects this would give 60% voids which is not that reasonable from the SEM images anymore. This suggests that inhomogeneous broadening caused by roughness (see figure 65) or deviations from short range order is dominating here, what would also explain the broader lines. Despite this derivations between experiment and calculation The spectral position of the two samples matches well with the predicted values. The transition from weak to strong coupling can be observed. While the sample from figure 66e) with $d_0/d_{coll}=1.59$ shows a clear blue shift compared to the spectrum of the isolated crescent (figure 66h)), the spectrum of the sample from figure 66a) with $d_0/d_{coll}=1.27$ does not show a blue shift but a red shift. This behaviour fits well to the simulation results in figure 63d) which predict a maximum blue shift at $d_0/d_{coll}=1.8$ and an increasing red shift for smaller values.

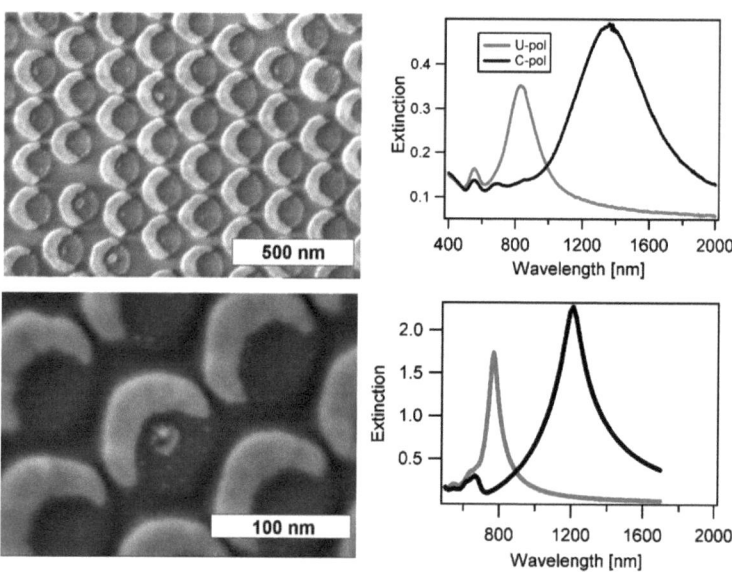

**Figure 67: Extinction spectra and representative scanning electron images of close packed crescents. The upper extinction spectrum is obtained experimentally while the lower one is calculated**

A sample where the masks have been reduced in size only slightly is shown in figure 67. Crescents are obtained with $d_{coll} = 185$ nm and $d_0 = 165$ nm as estimated from the electron micrographs. Even for the very small gaps of 20 nm between individuals a clear response in terms of dominating $U_1$ and $C_1$ peaks is seen. The resonances are again broader than predicted by the calculations. The $C_1$ resonance is much broader than the $U_1$ maximum both in the calculation and in the experiment. This can be seen as experimental verification of the resonance broadening due to strong inter-crescent coupling that is predicted by theory. Imperfections in the crescent shape and in the lattice lead to resonance broadening as well but, as observed for the arrays with larger crescent spacing this affects both $C_1$ and $U_1$ similarly.

One can conclude from this experiments together with the calculations, that the coupling of plasmonic resonators leads to two distinct coupling regimes when only subwavelength lattice spacings are considered. The first is a weak coupling regime at distances between half the wavelength and the penetration depth of the plasmonic resonances which results in small blue shifts and possibly a decreased line width. The second is the strong coupling regime where significant coupling between the near fields occur and results in strong red shifts and broadening of the resonances. While the latter regime is observable in experiments the former is strongly affected by imperfections of the fabricated samples and narrowing of the resonance line width could not be demonstrated experimentally.

## 8. Summary

In this thesis the sensing properties of plasmonic resonators for changes in refractive index were investigated. For the first time a self-consistent and general sensing theory was developed. This theory connects the electrodynamic properties of plasmonic resonators like resonance wavelength and electric field distribution to the sensitivity for refractive index changes. In measurements noise will lead to an uncertainty in the measured resonance frequency. This uncertainty can be expressed as a function of the signal-to-noise ratio (SNR) of the measurement and the quality factor of the resonance. A figure of merit (FOM) was derived which includes the effects of noise and in its general form directly states if a certain change in refractive index will be measurable or not. A remarkable fact of this FOM is that it includes the far field scattering properties like the scattering cross section or absorption cross section because these quantities enter the SNR of the measurement. This has not been considered before and as a consequence previously stated FOMs were incomplete. The derived FOM can be divided into a part which describes instrumental properties and one that describes resonator properties. It shows that particle and instrument optimization have to be considered equally and often instrument optimization will provide a greater optimization potential. For the resonator FOM in the quasi-static limit absolute bounds and scalings were derived. These bounds are based on the localization of electromagnetic energy for which analytic expressions have been derived before. The important result of the quasi-static considerations is that the sensitivity is completely determined by the choice of material and resonance wavelength for refractive index changes that cover the whole sensing volume. The sensitivity for smaller analytes can be expressed as this bulk limit times a weighting factor which describes the energy confinement to the analyte volume. This confinement has to be optimized together with the instrumental SNR in order to detect very small events like single molecules.

Optimization of the confinement can be done by introducing particles with sharp tips. Crescent shaped resonators are promising candidates for optimum resonators. Their sensing capabilities were investigated in this work. Based on the theoretical results numerical calculations and experiments were carried out. The numerical calculations confirm the developed theory. For an experimental verification, local refractive index changes were introduced close to the crescent shaped particles and their resonance wavelength change was measured. As a model analyte polystyrene colloids were used and manipulated with an atomic force microscope. This approach leads to a very defined and controllable model system with which the theoretical predictions could be verified without parasitic effects. The proposed theoretical model predicts the measured wavelength changes with high accuracy and allows to extrapolate the result to the single molecule response. The final result is that by increasing the SNR by a factor of eight it should be possible to detect the binding of single proteins with the investigated system.

The necessary instrument optimization remains to be done as well as the experimental detection of the binding and unbinding of a single molecule in solution. This thesis however shows how to systematically optimize the plasmonic resonators and opens the way to real single molecule sensitivity.

# Appendix A Measurement uncertainty for a Lorentzian peak

In this appendix the measurement uncertainty for the center frequency of a Lorentzian peak is derived. As shown in chapter 4.2 the measurement uncertainty is given by

$$\frac{\Delta}{N} \approx \frac{\Delta\omega_{un}^2}{\omega_{Det}} \int_{-\omega_{Det}/2}^{\omega_{Det}/2} \frac{1}{\sigma(\tilde{\omega})^2} \left( \frac{\partial SF(\tilde{\omega},\Gamma,\omega)}{\partial \omega} \bigg|_{\omega_c} \right)^2 d\tilde{\omega} \qquad (106)$$

The instrument function F is now a Lorentzian function

$$F = \frac{\Gamma^2}{4(\omega - \tilde{\omega})^2 + \Gamma^2} \qquad (107)$$

And the case of a Poissonian distribution of noise will be considered. In This case the quadratic error σ equals the signal. Inserting this in expression (106) we get

$$\frac{\Delta}{N} \approx \frac{\Delta\omega_{un}^2}{\omega_{Det}} \int_{-\omega_{Det}/2}^{\omega_{Det}/2} \frac{64S\Gamma^2(\omega-\omega_c)^2}{\left(\Gamma^2 + 4(\omega-\omega_c)^2\right)^3} d\tilde{\omega} \qquad (108)$$

As stated by Bobroff it is now convenient to do a coordinate transform

$$u = \frac{(\omega - \omega_c)}{\Gamma} \qquad (109)$$

And express the integration interval as the number of FWHMs t over which is integrated

$$\omega_{Det} = 2\Gamma t \qquad (110)$$

We arrive at

$$\frac{\Delta}{N} \approx \frac{\Delta\omega_{un}^2}{\Gamma^2} S \int_{-t}^{t} \frac{64u^2}{(1+4u^2)^3} du \qquad (111)$$

The integral is analytically solvable. The final result is

$$\Delta\omega_{un} \approx \frac{\Gamma}{\sqrt{S}} \sqrt{\frac{\Delta}{N}} \sqrt{\frac{1}{\frac{4t(4t^2-1)}{(4t^2+1)} + 2\arctan(2t)}} = \frac{\Gamma}{\sqrt{S}} \sqrt{\frac{\Delta}{N}} f(t) \qquad (112)$$

For the limit of infinity t $f = 1/\sqrt{\pi} \approx 0.564$ while for t=1 f=0.61. Now a level of confidence of 95% is assumed which gives for a three parameter fit Δ=4.9. Now Γ can be expressed in terms of the quality factor Q=ω/Γ. We arrive at an approximate expression for the relative uncertainty

$$\frac{\Delta\omega_{un}}{\omega} \approx \frac{1.3}{Q\sqrt{SN}} \tag{113}$$

For the wavelength the expression is the same

$$\frac{\Delta\lambda_{un}}{\lambda} \approx \frac{1.3}{Q\sqrt{SN}} \tag{114}$$

To have an approximate number of an uncertainty possible with a state of the art instrument a CCD spectrometer with 1024 pixels a 16 Bit AD converter and a Q of 10 is assumed. This gives $\sqrt{SN} = 2^{13} = 8192$ and an uncertainty of 0.002% or a wavelength accuracy of 0.02nm for a resonance wavelength of 800nm. It is very likely that this accuracy will not be reached in reality, due to drifts in the experimental setup (instrument noise) but the number gives a rough idea about the possibilities.

# Appendix B meep code

In this Appendix the source code of the meep FDTD simulation routines is shown and explained. Meep provides the basic time stepping routines as well as material properties and boundary conditions. It provides only electric and magnetic current sources and more advanced sources have to be built from this basic sources.

The setup of a simulation is done with a script written in the Scheme language. Please take a look at the meep homepage (ab-initio.mit.edu/meep) if you are not familiar with the way to do a simulation in meep. To make it easy to do parameter sweeps the simulation script is built very modular and controllable from the command line. Here are modules which a typical script consist of:

An Input section which contains definition and default values for all parameters that are made accessible from the command line.

A section in which inernally used parameters are defined

A section which prints information to the command line and an information file for each simulation

A section which contains geometry parameters for crescent shaped particles which are obtained from SEM images

A section where material models are defined

A section where geometry functions are defined. These functions are called with an vector argument and return the material as a function of position in space.

A section in which the simulation cell is defined. Here Geometry functions are called from

A section which defines the TFSF source

A section in which surfaces are defined over which the flux is integrated

A section which starts the simulation

Finally a section which prints simulation results to several files and the command line.

In the following a standard scheme file used to simulate a crescent shaped particle with an optional coating and/or attached sphere is shown. Usually this file is called from the command line with parameters that describe the actual simulation. For example the command

*mpirun.openmpi -n 4 meep-mpi res=400 sphereEp=2.5 spherePosPhi=70 Polarization=45 CrescentScript.scm*

calls the script with the distributed computing version of meep (with open mpi) on 4 processors a resolution of 400 pixels/μm and an incoming wave which is polarized 45° towards the crescent x-axis. Additionally there is an attached colloid with $\varepsilon=2.25$ at a position of 70°.

It should be possible to copy and paste this code and run it. The sections are clearly defined with comments (green text), and where necessary for understanding of the code extra comments are inserted. At some points the code looks very ugly due to additional line breaks when lines are longer in the original code file. To make it possible to copy and paste the code further formatting of this lines was not done. It is recommended to copy and paste the code to an ASCII file to study it.

```
;*******************************
;Input Section
;*******************************

    ;Geometry
    ;*****************
    (define-param OuterDiameter 0.205)  ; Outer Diameter of Crescent Geometry will get scaled for this value
    (define-param Thickness 0.04) ;Thickness of Crescent
    (define-param Theta 30) ; Evaporation Angle. Currently only 30 supported
    (define-param Phi 0) ; Turning Angle for double Evaporation.
    (define-param cutTips 0.0435) ;Position where the tip is cut This is at the moment use to simulation more realistic rounding of the tip but is really crap!
    (define-param coatT 0.00)

    (define-param attachSphere? true)
    (define-param sphereR 0.03)
    (define-param spherePosR (+ (/ OuterDiameter 2) (- sphereR 0.00)))
    (define-param spherePosPhi 50)
    (define-param sphereEp 1)


    ;Material
    ;*****************
    (define-param useGold? true) ; if false Silver will be used
    (define-param SilverModel1? true) ;True if one-oscillator Model, false if two-oscillator Model will be used

    ;Plane WaveSource
    ;*****************
    (define-param useGaussianSource? true);
    (define-param TFWidth 0.5) ;Width of totalfield area
    (define-param TFHeight 0.2) ;Height of total field area
    (define-param epsGlass 2.25) ; Permittiviy of Substrate
    (define centerFreq 0.9)  ; Mid frequency of pulse
```

```
(define pulseWidth 0.4) ;temporal width of pulse
(define-param Runtime 35)  ; RunTime for Continuos Source
(define-param Polarization 45) ; Angle of Crescent in degrees 0 is U-pol 90 is C_Pol
;TransmissionSpektrum
;********************
(define centerFreqFlux 1.15)  ; Mid frequency of monitored Frequency
(define fWidthFlux 1.1) ;Bandwith covered by the Monitor
(define-param nFreqs 100) ;Number of Frequencys for Transmission Spektrum

;Simulation Parameters
;*********************
(define-param normrun? false) ; True means simulation without Hoernchen
(set! eps-averaging? false)    ; Epsilon Averaging off. Makes no sense for Metals

;Computational Cell
;**********************
(define-param cellcross 0.8)       ;Transversal size of cell
(define-param cellheight 0.7)    ;cellheight of Cell (Direction of Propagation)
(define-param pmlthickness 0.1) ;Thickness of Absorbing Layers
(define-param pmlstrength 1);
(define-param res 200) ;Resulution in pixels per um

;Output
;************************
(define-param outputEField? false) ; Should the Efield be output ?
(define-param outputTimeStep 0.05) ; at what Step should Efield be output for Gaussian Source

(define-param outputTipField? false) ; Should The Field at The Tip be outbut into an extra File?
(define-param outputEpsilon? true)
(define-param OutputInterval 10)

;*****************************
;End Input Section
;*****************************


;*****************************
;Calculation of TFSF parameters
;*****************************

(set! Courant 0.5)  ; Courant number
(define pixel (/ 1.0 res))  ; size of a single pixel
(define dT (/ Courant res))  ; ; size of a time step
```

```
(define-param Ay (/ 1.0 res ));  Amplitude of H source on y sides


(define nGlass (sqrt epsGlass))
(define transA (/ (* 2 nGlass) (+ 1 nGlass)))  ; Fresnel Transmission coefficient glas/air
(define refA (/ (- nGlass 1) (+ 1 nGlass)))  ; Fresnel Reflection coefficient glas/air
; The following parameters control the delay between electric and magnetic currents they are maybe only necessary because of
a timing bug in meep. When electric and magnetic currents are not synchronized the TFSF source will leak energy into the SF
area which can be seen as a scattered flux even if no particle is present
(define-param n 1.0)  ;delay for magnetic source, in time steps
(define-param n2 0.3) ;delay for source on +z side
(define n3 1.0) ;delay for H source on y sides

(define nPix (* 8 (/ res 100)));How much Pixel to extend the source in the 2nd Medium
(define nPixMon (* 4 (/ res 100)));How much to shift the scattering monitor out and the the absorption
Monitor into the TF Area
; the nPixMon must be smaller than nPix otherwise the lower Boundary will be in the TF area

(define xPosTF (/ TFWidth 2))
(define zPosTF (/ TFHeight 2))
(define zCen (- (* nPix pixel) zPosTF ))

(define zsAbs  ( - TFHeight  (* 2 (* nPixMon pixel))))
(define zsScat (+ TFHeight (* 2 (* nPixMon pixel))))
(define xsAbs  ( - TFWidth   (* 2 (* nPixMon pixel))))
(define xsScat (+ TFWidth   (* 2 (* nPixMon pixel))))

(define xPosAbs  ( - (/ TFWidth 2) (* nPixMon pixel)))
(define xPosScat (+ (/ TFWidth 2) (* nPixMon pixel)))

;*****************************
;End Calculation of TFSF parameters
;*****************************


;*****************************
;Printing Stuff
;*****************************
(define (printinfo bla)
(print "\nWelcome to Andys Big Hoernchen Script!\n" )
(print "*****************************************\n" )
(print "\nInput parameters are: \n")
(print "\nGeometry Parameters are: \n")
(print " Outer Diameter = " OuterDiameter "\n")
(print " Thickness = " Thickness "\n")
```

```
(print " Theta = " Theta "\n")
(print " Phi = " Phi "\n")
(print " CutTips = " cutTips "\n")
(print " Coating Thickness = " coatT "\n")
(print " Sphere radius = " sphereR "\n")
(print " Spere phi = " spherePosPhi "\n")
(print " Sphere R = " spherePosR "\n")
(print " SphereEps= " sphereEp "\n")


(if useGold?
    (print " Material used is Gold \n")
    (if SilverModel1?
        (print " Material used is Silver1 \n")
        (print " Material used is Silver2 \n")
    )
)
(print "\nSource Parameters are: \n")
(if useGaussianSource?
    (print " Gaussian Source used \n Center Freqency = " centerFreq "\n With = " pulseWidth "\n")
    (print " Continous Source used \n Freqency = " centerFreq "\n Complex Fields are used!\n")
)
(print " Polarization: " Polarization "\n")
(print "\nCell Parameters are: \n")
(print " Crosssection of Computational Cell is " cellcross "\n")
(print " Height of Computational Cell is " cellheight "\n")
(print " PML thickness is " pmlthickness "\n")
(print " PML strength is " pmlstrength "\n")
(print " Resolution is " res "\n")
(print " Run time is " Runtime "\n")
(print "\n Output is:\n")
(if outputEField? (print " Efield is Stored "))
(if outputEField? (if useGaussianSource? (print "every " outputTimeStep "\n") (print "at end \n")))
(if outputTipField? (print "Tip Field is stored in extra File\n"))
(if outputEpsilon? (print " Epsilon is stored at Beginning\n\n"))
)
(printinfo 1)


;****************************
; End Printing Stuff
```

```
;********************************

(if (not useGaussianSource?) (set! force-complex-fields? true))
(set! progress-interval OutputInterval)

(if normrun? (set! filename-prefix (string-append "NormRun" "Res" (number-
>string res) "Pol" (number->string Polarization)))
    (set!   filename-prefix    (string-append    "CrescentD"    (number->string
OuterDiameter) "T" (number->string  Thickness) "Cut" (number->string cutTips)
"Coat" (number->string coatT) "SphereR" (number->string spherePosR) "SphereEps"
(number->string sphereEp) "SpherePhi" (number->string spherePosPhi) "Theta"
(number->string Theta) "Phi" (number->string Phi) "Res" (number->string res)
"Pol" (number->string Polarization)))
)


;************************************************
;Assigning Construction Parameters for Crescent
;
;The Crescent is constructed from a cylinder into which a shifted ellipse is cut
;The Values for the half axes and shift of the ellipse are derived from SEM pictures
;for each possible value of Theta and phi
;All values are taken from pictures made of Crescents with 150nm Colloids and are scaled
;to the actual radius of the cylinder
;************************************************

    (define scalingFactor (/ OuterDiameter 0.1))
    (define Cradius (* scalingFactor 0.05));Radius of Crescent-Circle

    (if  (= Phi 20) (print
                    (define ellip1 (* scalingFactor 0.0556));Radius of Elliptic SHadow1
                    (define ellip2 (* scalingFactor 0.0360));Radius of Elliptic SHadow2
                    (define middis (* scalingFactor 0.0373))


        )
    (if (= Phi 60)
                (print
                    (define ellip1 (* scalingFactor 0.0475));Radius of Elliptic SHadow1
                    (define ellip2 (* scalingFactor 0.0369));Radius of Elliptic SHadow2
                    (define middis (* scalingFactor 0.0225))


        )
     (if (= Phi 90)
                (print
```

```
                    (define ellip1 (* scalingFactor 0.0427));Radius of Elliptic SHadow1
                    (define ellip2 (* scalingFactor 0.032));Radius of Elliptic SHadow2
                    (define middis (* scalingFactor 0.0185))
      )
  (if (= Phi 120)
            (print
                    (define ellip1 (* scalingFactor 0.0433));Radius of Elliptic SHadow1
                    (define ellip2 (* scalingFactor 0.0316));Radius of Elliptic SHadow2
                    (define middis (* scalingFactor 0.0108))

      )

  (if (= Phi 0)
       (print
                    (define ellip1 (* scalingFactor 0.0508));Radius of Elliptic SHadow1
                    (define ellip2 (* scalingFactor 0.0413));Radius of Elliptic SHadow2
                    (define middis (* scalingFactor 0.0318))

       )
      )
     )
    )
 )

)


;Distance midCircle <--> midEllipse


(define thick Thickness)                      ;thickness of Structure
(define Hcellcross (+ (* 2 (+ Cradius coatT) ) (+ middis 0.01)))
(define TipVector (vector3 (* scalingFactor 0.035)
                           (* scalingFactor 0.036)
                           (/ thick 2)))      ;Position of a Tip of the Crescent


(define MonitorPositionPoint (vector3 0 0 (- (+ pmlthickness 0.05) (/ cellheight 2))))

(print "\n Debuginfos for Construction of Crescent \n")
```

```
(print "Crescent is build with the following parameters: \n")
(print "scaling Factor = " scalingFactor "\n")
(print "radius = " Cradius "\n")
(print "ellip1 = " ellip1 "\n")
(print "ellip2 = " ellip2 "\n")
(print "middis = " middis "\n")
(print "thick = " thick "\n")
(print "Hcellcross = " Hcellcross "\n")
(print "Tip Vector = " TipVector "\n")
(print "MonitorPositionPoint = " MonitorPositionPoint "\n")
```
;************************************************
;End Assigning Construction Parameters for Crescent
;************************************************


;************************************************
;Material Models
; These models are all derived by fitting of Lorentz oscillators to tabulated data.
;************************************************
;material model for Gold
;valid from lam=1.6um till lam=0.6um with max 4% in Realpart and 10% in Imagpart
```
(define gold (make dielectric (epsilon 9.166006)
              (polarizations
                  (make polarizability (omega 1e-20) (gamma 0.0570080) (delta-epsilon 52.255807e40))
                  (make polarizability (omega 1.650112) (gamma 0.0823894) (delta-epsilon 0.0683452))
              )
          )
)
```

;material model for Silver
;valid from lam=1.6um till lam=0.4um with max 3% in Realpart and 20% in Imagpart from 0.4 to 1um
;                         up to -40% in Imagpart from 1 to 1.6um
```
(define silver1 (make dielectric (epsilon 4.398687)
              (polarizations
                  (make polarizability (omega 1e-20) (gamma 0.00991936) (delta-epsilon 56.157452e40))
                  (make polarizability (omega 1.889370) (gamma 0.531923) (delta-epsilon 0.164729))
              )
          )
)
```

```
;material model for Silver with two Lorentz-Oscillators
;valid from lam=2um till lam=0.33um with max 10% in Realpart and 25% in Imagpart from 0.4 to 1um
(define silver2 (make dielectric (epsilon 4.663224)
                (polarizations
                    (make polarizability (omega 1e-20) (gamma 0.0106921) (delta-
epsilon 57.719524e40))
                    (make polarizability (omega 1.857922) (gamma 0.434866) (delta-
epsilon 0.125233))
             (make polarizability (omega 3.416286) (gamma 0.120041) (delta-epsilon
0.285034))
                )
            )
)

;Material Model for Glass
(define glass      (make dielectric (epsilon epsGlass)))
(define coat       (make dielectric (epsilon 2.25)))
(define sphereEps  (make dielectric (epsilon sphereEp)))
;**************************************************
;End Material Models
;**************************************************


;**************************************************
;Material Function for Crescent
; First is for a uncoated crescent while second also has a coating
;returns mat(x,y)
;**************************************************

(define (crescentfun pt)

    (set! pt (rotate-vector3 (vector3 0 0 1) (* Polarization (/ pi 180)) pt))

    (if (and (< (vector3-x pt) cutTips) (and
        (< (+ (* (vector3-x pt) (vector3-x pt)) (* (vector3-y pt ) (vector3-y
pt))) (* Cradius Cradius))
        (> (+ (/ (* (- (vector3-x pt) middis) (- (vector3-x pt) middis)) (*
ellip1 ellip1)) (/ (* (vector3-y pt ) (vector3-y pt)) (* ellip2 ellip2))) 1.0)))
                (if useGold? gold (if SilverModel1? silver1 silver2)) air)
)

(define (crescentfunCoating pt)

    (set! pt (rotate-vector3 (vector3 0 0 1) (* Polarization (/ pi 180)) pt))
```

```
        (if
        (and (< (vector3-x pt) (+ cutTips coatT))
             (and (not (and (> (vector3-x pt) middis) (< (abs (vector3-y pt)) (-
ellip2 coatT))))
                  (and
                        (< (+ (* (vector3-x pt) (vector3-x pt)) (* (vector3-y pt )
(vector3-y pt)))
                           (* (+ Cradius coatT) (+ Cradius coatT)))

                        (> (+ (/ (* (- (vector3-x pt) middis) (- (vector3-x pt) middis))
(* (- ellip1 coatT) (- ellip1 coatT))) (/ (* (vector3-y pt ) (vector3-y pt)) (*
(- ellip2 coatT) (- ellip2 coatT)))) 1.0)
                  )
            ))

        (if
           (and (> (vector3-z pt) (- thick)) (and (< (vector3-x pt) cutTips) (and
           (< (+ (* (vector3-x pt) (vector3-x pt)) (* (vector3-y pt ) (vector3-y
pt))) (* Cradius Cradius))
                 (> (+ (/ (* (- (vector3-x pt) middis) (- (vector3-x pt) middis)) (*
ellip1 ellip1)) (/ (* (vector3-y pt ) (vector3-y pt)) (* ellip2 ellip2)))
1.0))))

          (if useGold? gold (if SilverModel1? silver1 silver2)) coat)
    air)
)

;***********************************************
;End Material Function for Crescent
;***********************************************


;***********************************************
;Construction of the Simulation Cell
;***********************************************
;geometry input
(set! geometry-lattice (make lattice (size cellcross cellcross cellheight)))

   (if normrun?
    (set! geometry
        (list
            (make block (center 0 0 (/ cellheight 4)) (size cellcross cellcross
(/ cellheight 2))
```

```
                    (material glass))
  )
)

(set! geometry
    (list
      (make block (center 0 0 (/ cellheight 4)) (size 0.01 0.01 0.01)
                        (material gold)); this is just do initialize the dispersion function properly
```
which is otherwise not done due to a bug in meep. In newer versions this can be done due to initialistion of the material function earlier but this block has been kept to be compatible to old meep versions
                                          ; it is overwritten with the last object
```
      (make block (center 0 0 (- (/ (+ thick coatT) 2))) (size Hcellcross
Hcellcross (+ thick coatT))
                        (material (make material-function (material-func
crescentfunCoating))))
      (make block (center 0 0 (/ cellheight 4)) (size cellcross cellcross (/
cellheight 2))
                        (material glass))
      (make sphere (center (* spherePosR (cos (* (- spherePosPhi Polarization)
(/ pi 180))))
                                  (* spherePosR (sin (* (- spherePosPhi
Polarization) (/ pi 180)))) (- sphereR ))
                        (radius sphereR)
              (material sphereEps))
    )
   )
  )

  (set! pml-layers (list (make pml (thickness pmlthickness) (strength
pmlstrength))))

  (set! resolution res)
```

;************************************************
;End Construction of the Simulation Cell
;************************************************

;************************************************
; Sources
; This source construction describes a TFSF source which incorporates a wave reflected at a boundary between glass and air
; The wave is a differentiated Gaussian pulse (no DC) and travels in positive z-direction

; on the x/y sides the sources have to be switched on at the correct arrival time of the pulse started at -z
; The direct wave and transmitted and reflected waves are all treated separetely
;************************************************
(set! sources (list
                ;Incoming side
                (make source
                    (src    (make    gaussian-src    (frequency    centerFreq)    (width pulseWidth) (cutoff 5)))
                    (component Ex)
                    (center 0 0 (- (* nPix pixel) TFHeight))
                    (size TFWidth TFWidth 0))
                (make source
                    (src    (make    gaussian-src    (frequency    centerFreq)    (width pulseWidth) (cutoff 5) (start-time (* n dT)) ))
                    (amplitude 1) ;+1: emit to the left; -1: emit to the right.
            (component Hy)
                    (center 0 0 (- (* nPix pixel) TFHeight))
                    (size TFWidth TFWidth 0))
                ;outgoing side
            (make source
                    (src    (make    gaussian-src    (frequency    centerFreq)    (width pulseWidth) (cutoff 5)
                            (start-time   (+  (*  (*  nPix  pixel)  nGlass)  (+  (- TFHeight (* nPix pixel)) (* n2 dT)))))
                    (amplitude  (- transA)) ; pi phaseshift to the other side
            (component Ex)
                    (center 0 0 (+ (* pixel nPix)))
                    (size TFWidth TFWidth 0))
                (make source
                    (src    (make    gaussian-src    (frequency    centerFreq)    (width pulseWidth) (cutoff 5)
                            (start-time  (+  (*  (*  nPix  pixel)  nGlass)  (+  (- TFHeight (* nPix pixel)) (* (+ n n2) dT))))))
                    (amplitude  (- (/ transA nGlass)))  ; pi phaseshift to the other side
                    (component Hy)
                    (center 0 0 (+ (* pixel nPix)))
                    (size TFWidth TFWidth 0))
                ;Reflected Wave
                (make source
                    (src    (make    gaussian-src    (frequency    centerFreq)    (width pulseWidth) (cutoff 5)
                                (start-time (* 2 (- TFHeight (* nPix pixel))))))
                    (amplitude (+ refA))

```
        (component Ex)
                (center 0 0 (- (* nPix pixel) TFHeight))
                (size TFWidth TFWidth 0))
            (make source
                (src (make gaussian-src (frequency centerFreq) (width
pulseWidth) (cutoff 5)
                            (start-time (+ (* 2 (- TFHeight (* nPix pixel)))
(* n dT))) ))
                (amplitude (- refA))
        (component Hy)
                (center 0 0 (- (* nPix pixel) TFHeight))
                (size TFWidth TFWidth 0))
))

(do ((x (- (* ( + nPix 0) pixel) TFHeight) (+ x pixel))) ((> x 0))
;Sources at +-y sides
; these have to be startet each at a different time so that they are synchronous to the wave launched at the -x border
(print "build y source at z = " x " Start Time = " (+ x (- TFHeight (* nPix
pixel))) "\n")
(set! sources (append sources (list
                ;y Sides
        (make source
                (src (make gaussian-src (frequency centerFreq) (width
pulseWidth) (cutoff 5)
                            (start-time (+ (* n3 dT) (+ x (- TFHeight (*
nPix pixel)))))))
                (amplitude (- Ay))
        (component Hz)
                (center 0 (- 0 xPosTF) x)
                (size TFWidth 0))
        (make source
                (src (make gaussian-src (frequency centerFreq) (width
pulseWidth) (cutoff 5)
                            (start-time (+ (* n3 dT) (+ x (- TFHeight (*
nPix pixel)))))))
                (amplitude (+ Ay))
        (component Hz)
                (center 0 xPosTF x)
                (size TFWidth 0))
        ;reflected part
        (make source
                (src (make gaussian-src (frequency centerFreq) (width
pulseWidth) (cutoff 5)
```

```
                              (start-time  (+ (* n3 dT) (- (- TFHeight (* nPix
pixel)) x )))))
                (amplitude (+ (* Ay refA)))
        (component Hz)
                (center 0 (- 0 xPosTF) x)
                (size TFWidth 0))
        (make source
                (src  (make   gaussian-src  (frequency   centerFreq)   (width
pulseWidth) (cutoff 5)
                              (start-time  (+ (* n3 dT) (- (- TFHeight (* nPix
pixel)) x )))))
                (amplitude (- (* Ay refA)))
        (component Hz)
                (center 0 xPosTF x)
                (size TFWidth 0))
)
))
)

(do ((x (- (* ( + nPix 0) pixel) TFHeight) (+ x pixel))) ((> x 0))
(print "build x source at z = " x " Start Time = " (+ x (- TFHeight (* nPix
pixel))) "\n")
(set! sources (append sources (list
        ;xSides
        (make source
                (src  (make   gaussian-src  (frequency   centerFreq)   (width
pulseWidth) (cutoff 5)
                              (start-time  (+ (* 0 dT) (+ x (- TFHeight (*
nPix pixel)))))))
                (amplitude (- Ay))
        (component Ez)
                (center (- 0 xPosTF) 0 x)
                (size 0 TFWidth ))
        (make source
                (src  (make   gaussian-src  (frequency   centerFreq)   (width
pulseWidth) (cutoff 5)
                              (start-time  (+ (* 0 dT) (+ x (- TFHeight (* nPix
pixel)))))))
                (amplitude (+ Ay))
        (component Ez)
                (center xPosTF 0 x)
                (size 0 TFWidth ))
```

```
;reflected Part
        (make source
                (src    (make    gaussian-src    (frequency    centerFreq)    (width
pulseWidth) (cutoff 5)
                                (start-time (+ (* 0 dT) (- (- TFHeight (* nPix
pixel)) x ) ))))
                (amplitude (- (* Ay refA)))
         (component Ez)
                (center (- 0 xPosTF) 0 x)
                (size 0 TFWidth ))
        (make source
                (src    (make    gaussian-src    (frequency    centerFreq)    (width
pulseWidth) (cutoff 5)
                                (start-time (+ (* 0 dT) (- (- TFHeight (* nPix
pixel)) x ) ))))
                (amplitude (+ (* Ay refA)))
         (component Ez)
                (center xPosTF 0  x)
                (size 0 TFWidth ))
                )
))
)
;Transmitted Wave at Sides
(do ((x 0 (+ x pixel))) ((> (- x (* (- nPix 0) pixel)) 0))
;Sources at +-y sides
; these have to be startet each at a different time so that they are synchronous to the wave launched at the -x border
(print "build y source at z = " x " Start Time = " (+ (* 0 dT) (- TFHeight (*
nPix pixel)) (* x nGlass) ) "\n")
(set! sources (append sources (list
                ;y Sides
        (make source
                (src    (make    gaussian-src    (frequency    centerFreq)    (width
pulseWidth) (cutoff 5)
                                (start-time  (+ (* n3 dT) (- TFHeight (* nPix
pixel)) (* x nGlass) ))))
                (amplitude (- (* Ay (/ transA nGlass)) ))
         (component Hz)
                (center 0 (- 0 xPosTF) x)
                (size TFWidth 0))
        (make source
                (src    (make    gaussian-src    (frequency    centerFreq)    (width
pulseWidth) (cutoff 5)
```

```
                            (start-time   (+ (* n3 dT) (- TFHeight (* nPix
pixel)) (* x nGlass) ))))
                (amplitude (+ (* Ay (/ transA nGlass))))
        (component Hz)
                (center   xPosTF x)
                (size TFWidth  ))
)
))
)


(do ((x   (+ x pixel))) ((> (- x (* (- nPix  ) pixel))  ))
(print "build x source at z = " x " Start Time = " (+ (*   dT) (- TFHeight (*
nPix pixel)) (* x nGlass) ) "\n")
(set! sources (append sources (list
        ;xSides
        (make source
                (src  (make    gaussian-src   (frequency   centerFreq)   (width
pulseWidth) (cutoff  )
                                (start-time   (+ (*   dT) (- TFHeight (* nPix
pixel)) (* x nGlass) ))))
                (amplitude (- (* Ay transA)))
         (component Ez)
                (center (-   xPosTF)   x)
                (size   TFWidth ))
        (make source
                (src  (make    gaussian-src   (frequency   centerFreq)   (width
pulseWidth) (cutoff  )
                                (start-time   (+ (*   dT) (- TFHeight (* nPix
pixel)) (* x nGlass) ))))
                (amplitude (+ (* Ay transA)))
         (component Ez)
                (center xPosTF   x)
                (size   TFWidth ))
            )
))
)

;***********************************************
;End Sources
;***********************************************


;***********************************************
;Define Transmission Flux Object
```

;************************************************

```
(define fluxInside (add-flux centerFreqFlux fWidthFlux nFreqs ; Plane inside total Field Area
    (make flux-region (center 0 0 (- (* 5 pixel))) (direction Z) (size TFWidth
TFWidth)))
)


(define AbsInside ;Closed Surface inside total Field Area
 (add-flux centerFreqFlux fWidthFlux nFreqs
  (make flux-region (center xPosAbs 0 zCen)
                    (size 0 xsAbs zsAbs)
            (direction X)) ;+x

  (make flux-region (center (- xPosAbs) 0 zCen)
                    (size 0 xsAbs zsAbs ) (weight -1)
            (direction X)) ;-x

  (make flux-region (center 0 xPosAbs zCen)
                    (size xsAbs 0 zsAbs)
            (direction Y)) ;+y

  (make flux-region (center 0 (- xPosAbs) zCen)
                    (size xsAbs 0 zsAbs) (weight -1)
            (direction Y)) ;-y

  (make flux-region (center 0 0 (* (- nPix nPixMon) pixel) )
                    (size xsAbs xsAbs 0)
            (direction Z)) ;+z

  (make flux-region (center 0 0 (- (* (+ nPix nPixMon) pixel) TFHeight))
                    (size xsAbs xsAbs 0) (weight -1)
            (direction Z)) ;-z

 )
)

(define ScatOutside ;Closed Surface outside total Field Area
 (add-flux centerFreqFlux fWidthFlux nFreqs
  (make flux-region (center xPosScat 0 zCen)
                    (size 0 xsScat zsScat)
            (direction X)) ;+x
```

```
    (make flux-region (center (- xPosScat) 0 zCen)
                      (size 0 xsScat zsScat ) (weight -1)
              (direction X));-x

    (make flux-region (center 0 xPosScat zCen)
                      (size xsScat 0 zsScat)
              (direction Y));+y

    (make flux-region (center 0 (- xPosScat) zCen)
                      (size xsScat 0 zsScat) (weight -1)
              (direction Y));-y

    (make flux-region (center 0 0 (* (+ nPix nPixMon) pixel) )
                      (size xsScat xsScat 0)
              (direction Z));+z

    (make flux-region (center 0 0 (- (* (- nPix nPixMon) pixel) TFHeight))
                      (size xsScat xsScat 0) (weight -1)
              (direction Z));-z

 )
 )
```

;************************************************
;End Define Transmission Flux Object
;************************************************


;************************************************
; Run Function
;************************************************

```
    (run-until Runtime  ;(stop-when-fields-decayed 2 Ex MonitorPositionPoint 1e-6)
                (if  outputEpsilon?  (at-beginning  output-epsilon)(at-every  1e20
output-efield))
                (if outputEField?
                    (at-every outputTimeStep
                      (in-volume (volume (center 0 0 0 )
                         (size (- cellcross (* 2 pmlthickness))
                               (- cellcross (* 2 pmlthickness))
                               (- cellheight (* 2 pmlthickness))))))
```

```
                    output-efield-x output-efield-y output-efield-z
            )
          )
      (at-every 1e20 output-efield)
          )
            (if outputTipField? (to-appended "TipField.h5" (in-point TipVector
output-efield))
                          (at-every 1e20 output-efield))

  )


;*************************************************
;End Run Function
;*************************************************

;*************************************************
; Display Flux and Flux file out
;*************************************************
;(if useGaussianSource?
    (display-fluxes fluxInside ScatOutside AbsInside)
;)
(define defaultport (current-output-port))
(define    port1    (open-output-file   (string-append   filename-prefix
"TransmissionFluxSpectra.txt")))
(set-current-output-port port1)
;(if useGaussianSource?
    (display-fluxes fluxInside ScatOutside AbsInside)
;)
(set-current-output-port defaultport)
(close-output-port port1)
;*************************************************
;End Display Flux
;*************************************************


;*****************************
;Printing to file
;*****************************
(define defaultport (current-output-port))
(define  port1  (open-output-file  (string-append  filename-prefix  "Simulation
info.txt")))
(set-current-output-port port1)
(printinfo 1)
```

```
(print "\n Debuginfos for Construction of Crescent \n")
(print "Crescent is build with the following parameters: \n")
(print "scaling Factor = " scalingFactor "\n")
(print "radius = " Cradius "\n")
(print "ellip1 = " ellip1 "\n")
(print "ellip2 = " ellip2 "\n")
(print "middis = " middis "\n")
(print "thick = " thick "\n")
(print "Hcellcross = " Hcellcross "\n")
(print "Tip Vector = " TipVector "\n")
(print "MonitorPositionPoint = " MonitorPositionPoint "\n")
(print "Runtime = " Runtime "\n")

(set-current-output-port defaultport)
(close-output-port port1)
```

;*****************************
;Printing to file
;*****************************

# Appendix C Table of figures

Figure 1: Extinction for a 40nm in diameter gold sphere. .................................................. 7
Figure 2: Permittivity of silver. .................................................................................... 10
Figure 3: Permittivity of gold. ..................................................................................... 11
Figure 4: Partial functions of a sensor device ............................................................... 12
Figure 5: Basic confocal setup for reflection microscopy [22]. ...................................... 13
Figure 6: Annular illumination for dark field measurements [22] .................................. 14
Figure 7: Suppression of stray light in a confocal microscope [22] ............................... 14
Figure 8: Block diagram of partial functions and light signal flow of the microscope. . 15
Figure 9: Photography of the optimized confocal microscope. ...................................... 16
Figure 10: The arrangement of components of the confocal microscope ....................... 17
Figure 11: Typical scattering spectra from crescent shaped resonators .......................... 20
Figure 12: Integrated Intensity as a function of defocus (from [24]). ............................ 20
Figure 13: measured relative transmission of the confocal microscope (relative to t=0).21
Figure 14: Typical SEM image of crescents with colloids. ............................................ 22
Figure 15: SEM image of the evaporated TEM Grid and an additional scratch used for finding positions on the sample unambiguously. ................................................ 22
Figure 16: Close up of the grid.. .................................................................................. 23
Figure 17: The spatial lattice in the Yee algorithm on which E and H are evaluated [27].24
Figure 18: Construction of the computational cell with TFSF boundary and Monitors for the integrated total and scattered flux. .............................................................. 26
Figure 19: Superposition of electric and magnetic currents. .......................................... 27
Figure 20: 2D cut along the x-z plane of the three-dimensional simulation space. ........ 28
Figure 21: close-up of two discretized tips. .................................................................. 29
Figure 22: Dielectric function of gold from and fitted dielectric function .................... 30
Figure 23: Simulated spectra of a gold nanorod for different analytes.......................... 40
Figure 24: Convergence of the Integrals for the perturbation theory as a function of integration volume (given as the distance from the surface of the resonator). .................... 41
Figure 25: Convergence of the wavelengths shift from perturbation theory as a function of integration volume (given as the distance from the surface of the resonator). ....... 41
Figure 26: Extinction efficiencies for rods of different length. .................................... 42
Figure 27: Convergence of the Integrals for the perturbation theory as a function of integration volume (from the surface of the resonator) for a rod length 270 nm..................... 43

Figure 28: Convergence of the wavelengths shift from perturbation theory as a function of integration volume (from the surface of the resonator) for a rod length of 270nm.43

Figure 29: Convergence of the integrals for the perturbation theory as a function of integration volume (from the surface of the resonator for a rod length of 800 nm................... 44

Figure 30: Convergence of the wavelengths shift from perturbation theory as a function of integration volume (from the surface of the resonator) for a rod length of 800 nm..44

Figure 31: Transducers in different refractive index sensing concepts. ........................ 47

Figure 32: f(L) for a gold resonator. ............................................................................ 53

Figure 33: Sketch of the geometry of the crescent fabricable by nanosphere lithography on a glass support..................................................................................................................... 54

Figure 34: Sketch of the charge distribution of the first three resonances of a metal rod.56

Figure 35: Normalized peak shift, as calculated from the rod mode .............................. 58

Figure 36: Parametrisation of the rod ........................................................................ 59

Figure 37: Spectra of the bend rod for various bend radii. ........................................... 60

Figure 38: Resonance wavelength of the bend rod as a function of bending radius....... 60

Figure 39: Quality factor Q of the bend rod as a function of bending radius ................. 61

Figure 40: Extinction of a layer of isolated crescents .................................................... 63

Figure 41: calculated extinction spectra for 3 representative $t_{coat}$. ................................... 64

Figure 42: peak displacement as a function of coating thickness .................................. 65

Figure 43: peak displacement as a function of coating thickness .................................. 66

Figure 44: Response of the crescent as a function of sphere position ........................... 67

Figure 45: peak displacement for attachment of a colloid by perturbation theory.. ........ 68

Figure 46: peak displacement as a function of sphere radius ........................................ 68

Figure 47: Shifts of normalized extinction spectra decrasing crescent sizes. ................. 69

Figure 48: Resonance Wavelength and Q as a function of diameter for open crescents70

Figure 49: Resonance Wavelength and Q as a function of diameter for almost closed crescents 70

Figure 50: Field enhancement at a cut trough the middle of the particle in z direction for different sizes. .................................................................................................................. 71

Figure 51: Field enhancement at a cut trough the middle of the particle in z direction for different sizes .................................................................................................................. 72

Figure 52: relative peak shift for attachment of a dielectric colloid as a function of inverse crescent diameter................................................................................................................ 72

Figure 53: a) Calculated scattering spectra with and without colloid ............................ 74

Figure 54: Guiding Structure ..................................................................................... 76

Figure 55: a) Schematic representation of the experiment. ........................................... 78

Figure 56: Spectra of the crescent shaped particles before and after attachment of a PS colloid. 79

Figure 57: Histogram of spectral of shifts ........................................................................ 80

Figure 58: Spectra of the control structures that were not manipulated ......................... 81

Figure 59: Correlations between the parameters of the crescent and the peak displacement    82

Figure 60: simulated geometries of crescents with more realistic rounded tips and colloids attached at different positions.............................................................................................. 83

Figure 61: a, b) Sketches for two configurations which take into account a glass post of 20 nm in height below the crescent. .................................................................................... 84

Figure 62: Periodic Geometry considered for the calculations. ....................................... 87

Figure 63: a,b) Extinction spectra ($d_{coll}$=165 nm, t=40 nm, $\beta$=45°). ............................... 88

Figure 64: a) Field distribution at resonance for d0=500 nm and d0=185 nm for different relative orientations between the crescents and the underlying grating.............................. 90

Figure 65: Deformation of the contour ........................................................................... 90

Figure 66: Experimental results in the weak coupling regime......................................... 91

Figure 67: Extinction spectra and representative scanning electron images of close packed crescents. ............................................................................................................. 93

# Appendix D List of Symbols

| | |
|---|---|
| $A_\perp$ | Projected Area |
| $\alpha$ | Polarizability |
| $a_n$ | Expansion coefficients in Mie Theory |
| $\beta$ | Propagation constant |
| $b_n$ | Expansion coefficients in Mie Theory |
| C | Spatial energy confinement |
| C | Capacitance |
| c | Speed of light |
| $\chi$ | Susceptibility |
| $\chi^2$ | Sum of least squares |
| $\Delta$ | Confidence Level |
| $\Delta A$ | Detected signal |
| $\Delta A_{Un}$ | Uncertainty of a detected signal |
| $\Delta\varepsilon$ | Permittivity perturbation |
| $\Delta n$ | Refractive index perturbation |
| $\Delta\omega$ | Detected frequency shift |
| $\Delta\omega_{Un}$ | Uncertainty of detected frequency shift |
| $\vec{E}$ | Electric field strength |
| $\varepsilon$ | Relative permittivity |
| $\varepsilon_0$ | Electric constant |
| $\varepsilon_D$ | Dielectric permittivity |
| $\varepsilon_M$ | Metal permittivity |
| F | Spectral line shape |
| $\Phi$ | Scalar potential |
| $\phi$ | Phase |
| FOM | Figure of merit |
| $\Gamma$ | Full width at half maximum of a spectral line |
| $\vec{H}$ | Magnetic field strength |
| $H_n$ | Hankel function of order n |
| $I_{In}$ | Intensity of a plane wave |
| J | Electric current |
| k | Wave vector |
| $\kappa$ | Polarization vector |
| L | Depolarization factor |
| L | Inductance |

| | |
|---|---|
| $\lambda$ | Wavelength |
| $\lambda_0$ | Free space Wavelength |
| $l_d$ | Penetration depth |
| $\mu$ | Relative permeability |
| $\vec{M}$ | Magnetic current |
| m | Slope |
| $\mu_0$ | magnetic constant |
| $m_e$ | Electron mass |
| $\vec{M}_{eln}$ | Vector spherical harmonic |
| N | Number of detector channels |
| n | Refractive index |
| $n_0$ | Refractive index of surrounding |
| NA | Numerical aperture |
| $n_{eff}$ | Effective refractive index |
| $\vec{N}_{eln}$ | Vector spherical harmonic |
| P | Detector efficiency |
| $\vec{p}$ | Dipole moment |
| $P_{abs}$ | Absorbed Power |
| $P_{scat}$ | Scattered Power |
| Q | Quality factor |
| q | Electric charge |
| q | Electron charge |
| $Q_{abs}$ | Absorption efficiency |
| $Q_{scat}$ | Scattering efficiency |
| S | Detected Signal |
| $\sigma$ | Conductivity |
| $\sigma_{Abs}$ | Absorption cross section |
| $\sigma_{Ext}$ | Extinction cross section |
| SNR | Signal to noise ratio |
| $\sigma_{Scat}$ | Scattering cross section |
| T | Transmission |
| $\tau$ | Time |
| $\tau$ | Relaxation time |
| $U_D$ | Stored energy in dielectric |
| $U_M$ | Stored energy in metal |
| V | Electromagnetic Energy confined to analyte volume |
| V | Volume |
| $\omega$ | Angular frequency |
| $\omega_C$ | Center frequency of a resonance |

| | |
|---|---|
| $\omega_p$ | Plasma frequency |
| x | Displacement |
| z | Deviation from Focus |

# Literature

1. Homola, J., *Surface Plasmon Resonance Based Sensors.* 2006, New York: Springer.
2. Bohren, C.F. and D.R. Huffman, *Absorption and Scattering of Light by Small Particles.* 1998: Wiley-VCH, Berlin.
3. Kreibig, U. and M. Vollmer, *Optical Properties of Metal Clusters.* 1995, Berlin: Springer.
4. Englebienne, P., *Use of colloidal gold surface plasmon resonance peak shift to infer affinity constants from the interactions between protein antigens and antibodies specific for single or multiple epitopes.* Analyst, 1998. **123**(7): p. 1599-1603.
5. Raschke, G., *Molekulare Erkennung mit einzelnen Gold–Nanopartikeln.* 2005, Ludwig-Maximilian–Universität München: Munich.
6. Klar, T., et al., *Surface-plasmon resonances in single metallic nanoparticles.* Physical Review Letters, 1998. **80**(19): p. 4249-4252.
7. Sonnichsen, C., et al., *Spectroscopy of single metallic nanoparticles using total internal reflection microscopy.* Applied Physics Letters, 2000. **77**(19): p. 2949-2951.
8. Anker, J.N., et al., *Biosensing with plasmonic nanosensors.* Nature Materials, 2008. **7**(6): p. 442-453.
9. Mock, J.J., D.R. Smith, and S. Schultz, *Local refractive index dependence of plasmon resonance spectra from individual nanoparticles.* Nano Letters, 2003. **3**(4): p. 485-491.
10. Raschke, G., et al., *Biomolecular recognition based on single gold nanoparticle light scattering.* Nano Letters, 2003. **3**(7): p. 935-938.
11. Miller, M.M. and A.A. Lazarides, *Sensitivity of metal nanoparticle surface plasmon resonance to the dielectric environment.* Journal of Physical Chemistry B, 2005. **109**(46): p. 21556-21565.
12. Sherry, L.J., et al., *Localized surface plasmon resonance spectroscopy of single silver nanocubes.* Nano Letters, 2005. **5**(10): p. 2034-2038.
13. Rochholz, H., N. Bocchio, and M. Kreiter, *Tuning resonances on crescent-shaped noble-metal nanoparticles.* New Journal of Physics, 2007. **9**: p. 53
14. Shumaker-Parry, J.S., H. Rochholz, and M. Kreiter, *Fabrication of crescent-shaped optical antennas.* Advanced Materials, 2005. **17**(17): p. 2131-2134.
15. Mie, G., *Articles on the optical characteristics of turbid tubes, especially colloidal metal solutions.* Annalen Der Physik, 1908. **25**(3): p. 377-445.
16. Jackson, J.D. *Classical Electrodynamics.* [non-digital] 1998 [cited 2009; 3rd edition:[
17. Novotny, L. and B. Hecht, *Principles of Nano-optics.* 2006: Cambridge University Press.
18. Wang, F. and Y.R. Shen, *General properties of local plasmons in metal nanostructures.* Physical Review Letters, 2006. **97**(20).
19. Drude, P., *Zur Elektronentheorie der Metalle.* Annalen der Physik, 1900. **306**(3): p. 566-613.
20. Johnson, P.B. and R.W. Christy, *Optical-Constants of Noble-Metals.* Physical Review B, 1972. **6**(12): p. 4370-4379.
21. Genzel, L. and U. Kreibig, *Dielectric Function and Infrared-Absorption of Small Metal Particles.* Zeitschrift Fur Physik B-Condensed Matter, 1980. **37**(2): p. 93-101.
22. Genzel, L., T.P. Martin, and U. Kreibig, *Dielectric Function and Plasma Resonances of Small Metal Particles.* Zeitschrift Fur Physik B-Condensed Matter, 1975. **21**(4): p. 339-346.
23. Stefani, F.D., *Confocal Microscopy applied to the study of single entity flourescence and light scattering,* in *Chemistry.* 2005, Universität Mainz: Mainz.
24. Curry, A., et al., *Analysis of total uncertainty in spectral peak measurements for plasmonic nanoparticle-based biosensors.* Applied Optics, 2007. **46**(10): p. 1931-1939.

25. Wilson, T., R. Juskaitis, and P. Higdon, *The imaging of dielectric point scatterers in conventional and confocal polarisation microscopes.* Optics Communications, 1997. **141**(5-6): p. 298-313.
26. Taflove, A., *Computational Electromagnetics. The Finite Difference Time-Domain Method.* 1995: Artech House, Boston.
27. Oubre, C. and P. Nordlander, *Optical properties of metallodielectric nanostructures calculated using the finite difference time domain method.* Journal of Physical Chemistry B, 2004. **108**(46): p. 17740-17747.
28. Yee, K.S., *Numerical Solution of Initial Boundary Value Problems Involving Maxwells Equations in Isotropic Media.* Ieee Transactions on Antennas and Propagation, 1966. **AP14**(3): p. 302-&.
29. Farjadpour, A., et al., *Improving accuracy by subpixel smoothing in the finite-difference time domain.* Optics Letters, 2006. **31**(20): p. 2972-2974.
30. Tavlove, A., *Computational Electromagnetics. The Finite Difference Time-Domain Method.* 1995: Artech House, Boston.
31. Pomplun, J., et al., *Adaptive finite element method for simulation of optical nano structures.* Physica Status Solidi B-Basic Solid State Physics, 2007. **244**(10): p. 3419-3434.
32. Unger, A. and M. Kreiter, *Analyzing the Performance of Plasmonic Resonators for Dielectric Sensing.* J. Phys. Chem. C, 2009. **113**(28): p. 12243-12251.
33. Bocchio, N.L., et al., *Thin Layer Sensing with Multipolar Plasmonic Resonances.* J. Phys. Chem. C, 2008. **112**(37): p. 14355-14359.
34. Unger, A., et al., *Sensitivity of Crescent-Shaped Metal Nanoparticles to Attachment of Dielectric Colloids.* Nano Letters, 2009. **9**(6): p. 2311-2315.
35. Bobroff, N., *Position Measurement with a Resolution and Noise-Limited Instrument.* Review of Scientific Instruments, 1986. **57**(6): p. 1152-1157.
36. Bevington, P.R. and D.K. Robinson, *Data Reduction and Error Analysis for the Physical Sciences.* 1992, New York: McGraw-Hill.
37. Chew, W.C., *Waves and Fields in Inhomogeneous Media* IEEE Press Series on Electromagnetic Waves. 1999: IEEE Press.
38. Griffiths, D.J., *Introduction to Quantum Mechanics.* 1995, Englewood Cliffs, NJ: Prentice Hall.
39. Lai, H.M., et al., *Time-Independent Perturbation for Leaking Electromagnetic Modes in Open Systems with Application to Resonances in Microdroplets.* Physical Review A, 1990. **41**(9): p. 5187-5198.
40. Joannopoulos, J.D.a.J., Steven G. and Winn, Joshua N. and Meade, Robert D., *Photonic Crystals: Molding the Flow of Light (Second Edition).* 2008: Princeton University Press.
41. Landau, L.D. and E.M. Lifshitz, *Lehrbuch der theoretischen Physik VIII Elektrodynamik der Kontinua.* Vol. VIII. 1985, Berlin: Akademie Verlag
42. Armani, A.M., et al., *Label-free, single-molecule detection with optical microcavities.* Science, 2007. **317**(5839): p. 783-787.
43. Arnold, S., et al., *Shift of whispering-gallery modes in microspheres by protein adsorption.* Optics Letters, 2003. **28**(4): p. 272-274.
44. Chow, E., et al., *Ultracompact biochemical sensor built with two-dimensional photonic crystal microcavity.* Optics Letters, 2004. **29**(10): p. 1093-1095.
45. Dostalek, J. and W. Knoll, *Biosensors based on surface plasmon-enhanced fluorescence spectroscopy.* Biointerphases, 2008. **3**(3): p. FD12-FD22.
46. Bergman, D.J. and M.I. Stockman, *Surface plasmon amplification by stimulated emission of radiation: Quantum generation of coherent surface plasmons in nanosystems.* Physical Review Letters, 2003. **90**(2).
47. Noginov, M.A., et al., *Demonstration of a spaser-based nanolaser.* Nature, 2009. **460**(7259): p. 1110-U68.

48. Novotny, L., *Effective wavelength scaling for optical antennas.* Physical Review Letters, 2007. **98**(26).
49. Buckman, B.A., *Guided wave Photonics*, ed. S.C.H.B. Publishing. 1995: Saunders College/ Harcourt Brace Publishing
50. Zhou, J., et al., *Saturation of the magnetic response of split-ring resonators at optical frequencies.* Physical Review Letters, 2005. **95**(22).
51. Feigenbaum, E. and M. Orenstein, *Ultrasmall Volume Plasmons, yet with Complete Retardation Effects.* Physical Review Letters, 2008. **101**(16).
52. Nusz, G.J., et al., *Rational Selection of Gold Nanorod Geometry for Label-Free Plasmonic Biosensors.* ACS Nano, 2009. **3**(4): p. 795-806.
53. Voros, J., *The density and refractive index of adsorbing protein layers.* Biophysical Journal, 2004. **87**(1): p. 553-561.
54. Larsson, E.M., et al., *Sensing characteristics of NIR localized surface plasmon resonances in gold nanorings for application as ultrasensitive biosensors.* Nano Letters, 2007. **7**(5): p. 1256-1263.
55. Stiles, R.L., et al., *Investigating tip-nanoparticle interactions in spatially correlated total internal reflection plasmon spectroscopy and atomic force microscopy.* Journal of Physical Chemistry C, 2008. **112**(31): p. 11696-11701.
56. Fisher, L.R. and J.N. Israelachvili, *Experimental Studies on the Applicability of the Kelvin Equation to Highly Curved Concave Menisci.* Journal of Colloid and Interface Science, 1981. **80**(2): p. 528-541.
57. Greaves, E.D., et al., *Optimizing accuracy and precision in spectral data.* X-Ray Spectrometry, 2005. **34**(3): p. 194-199.
58. Markus, R., et al., *Parallel Preparation of Densely Packed Arrays of 150-nm Gold-Nanocrescent Resonators in Three Dimensions.* Small, 2009. **5**(18): p. 2105-2110.
59. Wood, R.W., *On a remarkable case of uneven distribution of light in a diffraction grating spectrum.* Philosophical Magazine, 1902. **4**(19-24): p. 396-402.
60. Carron, K.T., et al., *Resonances Of Two-Dimensional Particle Gratings In Surface-Enhanced Raman-Scattering.* Journal Of The Optical Society Of America B-Optical Physics, 1986. **3**(3): p. 430-440.
61. de Abajo, F.J.G., *Colloquium: Light scattering by particle and hole arrays.* Reviews of Modern Physics, 2007. **79**(4): p. 1267-1290.
62. Haes, A.J., et al., *A nanoscale optical biosensor: The long range distance dependence of the localized surface plasmon resonance of noble metal nanoparticles.* Journal of Physical Chemistry B, 2004. **108**(1): p. 109-116.
63. Zou, S.L. and G.C. Schatz, *Theoretical studies of plasmon resonances in one-dimensional nanoparticle chains: narrow lineshapes with tunable widths.* Nanotechnology, 2006. **17**(11): p. 2813-2820.
64. Auguie, B. and W.L. Barnes, *Collective resonances in gold nanoparticle arrays.* Physical Review Letters, 2008. **101**(14).
65. Auguie, B. and W.L. Barnes, *Diffractive coupling in gold nanoparticle arrays and the effect of disorder.* Optics Letters, 2009. **34**(4): p. 401-403.
66. Felidj, N., et al., *Grating-induced plasmon mode in gold nanoparticle arrays.* Journal Of Chemical Physics, 2005. **123**(22).
67. Haynes, C.L., et al., *Nanoparticle optics: The importance of radiative dipole coupling in two-dimensional nanoparticle arrays.* Journal of Physical Chemistry B, 2003. **107**(30): p. 7337-7342.
68. Sung, J., et al., *Nanoparticle spectroscopy: Dipole coupling in two-dimensional arrays of L-shaped silver nanoparticles.* Journal Of Physical Chemistry C, 2007. **111**(28): p. 10368-10376.

# List of Publications

**Refereed Journals**

Unger, A. and M. Kreiter, *Analyzing the Performance of Plasmonic Resonators for Dielectric Sensing.* J. Phys. Chem. C, 2009. **113**(28): p. 12243-12251.

Unger, A., et al., *Sensitivity of Crescent-Shaped Metal Nanoparticles to Attachment of Dielectric Colloids.* Nano Letters, 2009. **9**(6): p. 2311-2315.

Bocchio, N.L., Unger, A. et al., *Thin Layer Sensing with Multipolar Plasmonic Resonances.* J. Phys. Chem. C, 2008. **112**(37): p. 14355-14359.

**Invited Talks**

A. Unger, M. Kreiter, Optimization of nanoparticles for label free detection of single molecules via plasmonic resonances, EPFL 2009 Group of Prof. Dr. T. Lasser

**Conference Contributions**

M. Kreiter, A. Unger, N. Bocchio, U. Rietzler, R. Berger, How to find the optimum plasmonic resonator for the sensing of single biomolecules and dielectric layers? Molecular Plasmonics 2009 (talk)

A. Unger, M. Kreiter, K.-H. Brenner, Optimization of nanoparticles for label free detection of single molecules via plasmonic resonances DGAO Proceedings 2009 (talk)

A. Unger, M. Kreiter, Analyzing the Performance of Plasmonic Resonators for Dielectric Sensing, TaCoNa-Photonics 2009 (poster)

A. Unger, K.-H. Brenner, Vergleich exakter optischer Lösungsmethoden im Zeitbereich in Hinblick auf Genauigkeit und Effizienz, DGAO Proceedings 2008 (talk)

A. Unger, M. Kreiter, S. Burger, Simulation of crescent shaped nanoresonators for biosensing applications TaCoNa-Photonics 2008 (poster)

Die VDM Verlagsservicegesellschaft sucht für wissenschaftliche Verlage abgeschlossene und herausragende

## Dissertationen, Habilitationen, Diplomarbeiten, Master Theses, Magisterarbeiten usw.

für die kostenlose Publikation als Fachbuch.

Sie verfügen über eine Arbeit, die hohen inhaltlichen und formalen Ansprüchen genügt, und haben Interesse an einer honorarvergüteten Publikation?

Dann senden Sie bitte erste Informationen über sich und Ihre Arbeit per Email an *info@vdm-vsg.de*.

**Sie erhalten kurzfristig unser Feedback!**

VDM Verlagsservicegesellschaft mbH
Dudweiler Landstr. 99
D - 66123 Saarbrücken
**www.vdm-vsg.de**

Telefon +49 681 3720 174
Fax     +49 681 3720 1749

Die VDM Verlagsservicegesellschaft mbH vertritt

Printed by Books on Demand GmbH, Norderstedt / Germany